W9-BTL-131

Masters of Modern Physics

Published Volumes

The Road from Los Alamos by Hans A. Bethe
The Charm of Physics by Sheldon L. Glashow
Citizen Scientist by Frank von Hippel

THE CHARM OF PHYSICS

SHELDON L. GLASHOW

A TOUCHSTONE BOOK
Published by Simon & Schuster
New York London Toronto Sydney Tokyo Singapore

Touchstone
Simon & Schuster Building
Rockefeller Center
1230 Avenue of the Americas
New York, New York 10020

First Touchstone Edition 1991
Published by arrangement with The American Institute of Physics

TOUCHSTONE and colophon are registered trademarks
of Simon & Schuster Inc.

Manufactured in the United States of America

10 9 8 7 6 5 4 3 2 Pbk.

ISBN 0-671-74013-X Pbk.

This book is volume one of the *Masters of Modern Physics* series

Contents

Series Introduction vii

Preface ix

THE LIFE OF A PHYSICIST

Elementary-Particle Physics and Me 3

Internal Exile in California 10

The Mysteries of Matter 16
with Peter Costa

THE WORLD OF SCIENCE

A Peek at the Universe 27

Life on Log Time 42

The Number Game 55

Welcome to UBS 59

Are We Alone in the Universe? 74

The Big Picture 93

THE WORK OF A THEORIST: ELEMENTARY PARTICLES

What Is an Elementary Particle? 109

The Hunting of the Quark 126

Quarks with Color and Flavor 141

The Invention and Discovery of the Charmed Quark 164

Antineutrinos and Geology 168

Elementary-Particle Physics as a Waste of Time and Money 171

Does Elementary-Particle Physics Have a Future? 179

Big Things, Little Things 192

THE WORK OF A THEORIST: GRAND UNIFICATION

Towards a Unified Theory: Threads in a Tapestry 203

Unified Theory of Elementary-Particle Forces 214
with Howard Georgi

Grand Unification 239

On the Way to a Unified Field Theory 248

Tangled in Superstring 252
with Ben Bova

Grand Unification and the Future of Physics 257

THE PHYSICIST AND SOCIETY

SSC: Machine for the Nineties 265
with Leon M. Lederman

Passing the Torch 285

Teaching the Lowest Common Denominator 288

Science and Violence 291

Acknowledgments 295

Index 301

About the Author

Series Introduction

In his pessimistic essay, *The Two Cultures,* published in the late 1950s, C.P. Snow lamented that the world occupied by scientists and the world in which the rest of us live were rapidly moving apart. The arcane rites practiced by technical people were becoming incomprehensible, even to scholars who knew something about Shakespeare or Stendahl. Snow believed that as they became more inaccessible to each other, the divided cultures would never speak to each other again.

But in the next few years, the two cultures were thrown together in ways that Snow had not anticipated. The products of science and technology intruded themselves into our daily lives, making it impossible to live in the modern world without tripping over the connections. Science is no longer secreted away in remote laboratories where apparatus bubbles in the night. It is everywhere.

The automobile was the first to invade. With it came not only the social transformation of leisure and work, but it introduced the complexities of fuels, combustion, and chemistry into everyday life. Now the expert was not the only one who knew something. The kid down the block could tell you a thing or two about how cars worked too—even if he knew nothing of Greek mythology or relativity.

Today, we're all so familiar with airplanes and helicopters, with VCRs and computers, with contraceptives and mood-altering drugs, they no longer appear to us as having been created by science and technology. They now take their ordinary places next to your box of breakfast cereal.

We now talk with casual familiarity about our cholesterol levels or about the ecological consequences of ozone holes. Even kids adopt techno-jargon as street slang—blast-off, nuke, clone. People are much more familiar with Stephen Hawking than they are with Stephen Daedalus, even though Stephen King still wins that popularity contest.

That is not to say that the untrained know the foundations of physics or the concepts that underlie other sciences. When pushed, however, people do find out things. Those who first introduced nuclear power onto the American landscape in the fifties discovered to their surprise that large numbers of technically unsophisticated citizens could learn a great deal about reactors. Similarly, ordinary people have educated themselves about other technologies that play a central role in our environment or that have consequences elsewhere—drugs, abortion, AIDS.

As science and technology become more intricately threaded into our existence, the need for us to understand what it is all about, how it works, and what it does becomes very personal. As the turn of the century approaches, we will no longer be able to look for jobs merely with the strength of our hands or our native intelligence. The intimate way you live your life—the food you eat, the fuels you burn at home or in your car, even how you choose a partner—will depend in part on your technical sophistication.

For these and other, broader cultural reasons, the Masters of Modern Physics series has been launched by the American Institute of Physics, a consortium of major physics societies. The series introduces to the reading public the work and thought of some of the most celebrated physicists of our day. These volumes of collected essays offer a panoramic tour of the way science works, how it affects our lives, and what it means to those who practice it. Authors report from the horizons of modern research, provide engaging sketches of friends and colleagues, and reflect on the social, economic, and political consequences of the scientific and technical enterprise.

Authors have been selected for their contributions to science and for their keen ability to communicate to the general reader—often with wit, frequently in fine literary style. All have been honored by their peers and most have been prominent in shaping debates in science, technology, and public policy. Some have achieved distinction in social and cultural spheres outside the laboratory.

Many essays are drawn from popular and scientific magazines, newspapers, and journals. Still others—written for the series or drawn from notes for other occasions—appear for the first time. Authors have provided introductions and, where appropriate, annotations. Once selected for inclusion, the essays are carefully edited and updated so that each volume emerges as a finely shaped work.

Robert N. Ubell

Preface

As a small child, I was fascinated by the mechanisms in things such as clocks and radios, and felt compelled to take them apart to see how they worked. My children are far less destructive: they are not likely to become scientists. The Second World War impelled me to choose science as a career. When I was ten years old, my elder brother (who would later be a glider trooper in Africa and Europe, and is now a retired dentist) explained to me why low-flying aircraft had to take evasive action after releasing their bombs—that was my introduction to classical mechanics. A few years later, two bombs were dropped on Japan. I couldn't take a bomb apart, but I had to try to understand news references to neutrons, binding energies and chain reactions. Articles in *Astounding Science Fiction* were my introduction to relativity and quantum theory.

After the War, my father installed a small chemistry laboratory for me in the basement of our house in Manhattan. I had outgrown chemistry sets, and had become intrigued (who knows why?) by the chemistry of selenium. My "research project"—the substitution of selenium for sulfur in the hydroculture of tomato seedlings—led to my selection as a finalist in the Westinghouse Science Talent Search. I began preparing for my career as a physicist as an undergraduate at Cornell University, and subsequently, as a graduate student at Harvard. A capsule biography, which I recently wrote for the 1989 Harvard-Radcliffe Yearbook, summarizes what happened next:

> When I came to Harvard in 1954, I discovered, to my horror, that lowly graduate students had nothing like Cornell's "Game Room" to hang out in. Instead, I was soon installed in a windowless warren of tiny cubicles

in the dank cellar of Lyman Laboratory. Here, assisted by an array of antique and noisy mechanical calculators, a collection of worn and worthless books, occasionally encouraged by an avuncular nod from my research supervisor Julian Schwinger, I was expected to generate a monumental contribution to Science—with six copies, typed, printed, bound, and on 100 percent (or else!) rag paper. My pet leech offered scant solace.

Four years later I escaped to enjoy a two-year paid European holiday courtesy of the National Science Foundation. [Note added:—Perhaps it was not all vacation. During those years I did the work that would earn me a Nobel Prize two decades later.] My thesis had been duly buried in the Harvard Archives, my leech under the World Tree. While I was abroad, Murray Gell-Mann, the hyphenated guru of elementary particles, mistook me for an East German prodigy and lured me to Cal Tech for what would become a 66-month period of exile in a strange land. I discovered smog, was ticketed for jay-walking, played go, went rock climbing, taught at Stanford and Berkeley, and survived the Filthy Speech Movement.

In tenure and triumph, I returned to Harvard in 1966 with a mere thirty percent cut in salary. Now at last, I could do unto others what Harvard had done unto me! There was still no tolerable pool room in Cambridge, nor is there likely to be one with today's no-smoking ordinances. So it's back to work on the one problem that has always fascinated me—that has passed through the minds of generations of scientists, that was entrusted to me by Schwinger, and that I must entrust to my students to address in the inimitable fashion of the physicist—''What does it all mean?''

Implicit in this tongue-in-cheek essay is the theme of this volume: physics can be just plain fun as well as an obsession. How marvellous it is that we are paid for what we so love to do!

What changes have been wrought since I started my career! Men on the moon, pocket calculators, VCRs—the whole face of science has changed with the emergence of new forefront disciplines such as recombinant genetics and plate tectonics. But the most radical transformations have taken place in particle physics and cosmology, opposite ends on a scale of sizes. When I began my studies, scientists had no firm notion of how the universe began or what an elementary particle is, although in recent years two ''standard models'' have been developed: one of Big Bang cosmology, the other of particle physics in terms of quantum chromodynamics and the electroweak model. The macro- and the micro-theories are compatible, both are supported by an impressive body of empirical evidence, and no established data has yet contradicted either theory. Together, these two structures constitute the beginning of a formal theory underlying all there is, was, and will be. It is almost a religion, with its own Genesis (. . . and then, 15 billion years ago, there was light!), and its own Trinity (the proton: indivisible, but made of three quarks). But we are not there yet. The toughest questions remain unanswered.

How did stars group together in galaxies, which themselves are gathered in clusters and even in clusters of clusters, all of which surround vast and empty spherical voids patterned rather like suds in a kitchen sink? We do not know what constitutes the very matter of the Universe, for stars, clouds of gas, and dust cannot account for its mass. Although most matter is invisible to us, we are aware of its gravitational effects. From the index of 17 fundamental particles, astronomers have (reluctantly, to be sure) driven us to conclude that the universe is predominantly made of something eerie and unearthly, which is almost certainly "none of the above."

We do not know why particles exist, why they have a particular mass, or why they are subject to certain forces. Our standard model is honest: it tells us that within this context, there are no answers. Moreover, our "grand unified theory" is neither grand (the proton lifetime is wrong), nor unified (gravity is left out), nor is it really a theory (it doesn't solve any of the above puzzles).

Quantum field theory, the marriage of quantum mechanics and special relativity as devised by Dirac, Schwinger, Feynman, Tomonaga, and others, has served us well for over 40 years, and helped establish our standard model of particle physics. But it has reached an impasse; the theory simply cannot describe gravity, and is therefore unable to explain the earliest moments of the creation of the universe. Quantum field theory also falls short of answering any of today's particle puzzles. Evidently, we need to make a giant step toward a far more powerful conceptual framework. Superstring theory is an ambitious and fashionable attempt to overcome all obstacles by supposing that elementary particles are actually tiny loops of string, rather than pointlike structures. Within the context of superstrings, a quantum theory of gravity emerges naturally. Perhaps someday a theory will be able to answer Isadore Rabi's famous query: Who needs the muon? So far, superstring theory cannot begin to answer the question, but has rather led a whole generation of brilliant graduate students into an increasingly intricate ten-dimensional mathematical morass.

At the 1989 conclave of the International Summer School on Subnuclear Physics in Erice, Antonino Zichichi, its founder and director, asked one of his younger and more abstract speakers, "Can any experiment, or even an imagined experiment, determine whether what you have said is true?'' The answer was negative. Zichichi's question is of prime importance in this age where mathematics can masquerade as physics. Julian Schwinger believes in a unified theory of everything, but he (and I) feel that the time is not yet ripe for such hubris.

Particle physicists and cosmologists—who operate at opposite extremes on the cosmic ladder—are among the last prophets of a scientific revolution. We hope for data that will invalidate the theoretical framework we have so painstakingly built. We look forward to a time when the essays in this collection become obsolete because our successors will have formulated a better theory of how things work and what it all means. The future will be chock-full of experimental surprises that will compel us to revamp and improve our standard models and approach more closely the One and True Synthesis.

THE LIFE OF A PHYSICIST

Elementary-Particle Physics and Me

My parents, Lewis Glashow and Bella née Rubin immigrated to New York City from Bobruisk in the early years of this century. In the U.S. they found the freedom and opportunity denied Jews in Czarist Russia. After years of struggle, my father became a successful plumber, and his family could then enjoy the comforts of the middle class. While my parents never had the time or money to secure university education themselves, they were adamant that their children should. In comfort and in love, we were taught the joys of knowledge and of work well done. I only regret that neither my mother nor my father could live to see the day I would accept the Nobel Prize.

When I was born in Manhattan in 1932, my brothers Samuel and Jules were 18 and 14 years old. They chose careers of dentistry and medicine, to my parents' satisfaction. From an early age, I knew I would become a scientist. It may have been my brother Sam's doing. He interested me in the laws of falling bodies when I was 10, and helped my father equip a basement chemistry lab for me when I was 15. I became skilled in the synthesis of selenium halides. Never again would I do such dangerous research. Except for the occasional suggestion that I should become a physician and do science in my spare time, my parents always encouraged my scientific inclinations.

Among my chums at the Bronx High School of Science were Gary Feinberg and Steven Weinberg. We spurred one another to learn physics while commuting on the New York subway. Another classmate, Dan Greenberger, taught me calculus in the school lunchroom. High-school mathematics then terminated with solid geometry. At Cornell University,

I again had the good fortune to join a talented class. It included the mathematician Daniel Kleitman, my old classmate Steven Weinberg, and many others who were to become prominent scientists. Throughout my formal education, I would learn as much from my peers as from my teachers. So it is today among our graduate students.

I came to Harvard as a graduate student of physics in the fall of 1954. Danny Kleitman, Dave Falk, and I were the Cornell contingent. Steve Weinberg, Laurence Mittag, and Tema Ehrenreich were also of the notorious Cornell crowd. Steve chose Princeton, by which I was rejected; Laurence chose Yale and now teaches engineers at Boston University; Tema (Mrs. Henry Ehrenreich) is part of the history of condensed matter physics at Harvard. During my first year, I collaborated on a paper about nuclear physics with Walter Selove, then a junior faculty member. It was my only sortie into this antique and exotic discipline. Today, Harvard students of the nuclear persuasion must cross-register at MIT or, preferably, switch.

Mostly, I took a lot of courses, including especially the inspired and inspiring lectures of Julian Schwinger, my thesis director-to-be. The late Jun Sakurai, as a precocious Harvard senior was a classmate, as was T.T. Wu, the noted particle physics presence. Among the other courses was John Van Vleck's famed group theory. Over the years, it has evolved into an entirely unrecognizable but absolutely essential course in group theory for particle physics, now taught by Howard Georgi. I also endured the first course ever taught by Paul Martin—relativity: special, general, and incomprehensible.

Requirements were to be met in two foreign languages, and in what for me was yet a third, laboratory work. French was a pushover, but my passing the Russian exam was a mean trick. There is no longer any foreign language requirement in our department. Physics, the world over, is done in Pidgin English. A graduate version of the college's expository writing course might be a useful requirement today.

Since I had already experienced the joys and sorrows of dropping lead bricks on fragile geiger tubes at Cornell, I had no intention of doing yet another laboratory course. I exercised my option to take an extra oral examination in methods of experimental physics. Ken Bainbridge and Curry Street were my examiners. I passed. In the intervening decades, to my knowledge, no one else has chosen this route of avoiding the lab course.

The time had come to choose my thesis director. About 10 of us approached Professor Schwinger simultaneously: Kleitman, Falk, Marshal Baker, Charlie Sommerfield, Harold Weitzner, Ray Sawyer, Charlie Warner, Bob Warnock, me, and perhaps another. The master assigned us the problem of computing the electron propagator in Coulomb gauge. Collaborating, we appeared a few days later, en masse, with the solution. Schwinger's strategy had failed; he had to invent a different problem for each of us.

Charlie Sommerfield got the problem of how an electron is affected by a magnetic field. He was to show that the original fourth-order calculation of Karplus and Kroll was flawed. Eventually, he was the first to compute the anomalous moment to sixth order. Now at Yale, he leaves the question of yet higher corrections to others.

Danny Kleitman, as I recall, was assigned the properties of an electron in an electric field. Eventually, he completed his doctoral research partly with Schwinger and partly with Roy Glauber. One of Danny's accomplishments was the derivation of the analog to the Gell-Mann–Okubo mass formula in Schwinger's global symmetry scheme. The formula doesn't work, and global symmetry is all wet. Danny has since become a great mathematician, and my brother-in-law as well.

I was to work on the possible synthesis of weak and electromagnetic interactions by means of the newly invented Yang-Mills, or non-Abelian, gauge theories. It was a bit premature. The experimentalists had not yet identified the VA form of weak interactions. The old dogma (STP) based upon a silly theory and three incorrect experiments, had only recently given way to VT. One false experiment had not yet been exorcised, and furthermore, parity violation had not yet been discovered.

During my years at Harvard, I particularly remember interactions with Wally Gilbert, Chuck Zemach, Ken Johnson, Irwin Shapiro, George Sudarshan, and Abe Klein. From Abe, all I wanted was a reading course in field theory. ''Junior Fellows are not obliged to teach,'' more or less, was Klein's comment. Irwin was Roy Glauber's first student. He maintained a sort-of salon on Shepard Street, and gave many large parties. Chuck was privy to the last and unpublished portion of Schwinger's great opus: TQF, or the theory of quantum fields. Unfortunately, I could never understand it. In my third year at graduate school, Chuck and I shared an apartment in Brighton. We gave at least one glorious party. Nodding off early in the evening, I was to discover, several days later, that Roy Glauber had put a frankfurter in my flute. Wally Gilbert, at Harvard, was completing a thesis under Salam at Imperial College. It was something about backwards dispersion relations, as I remember. He was, by far, the most mathematical of physicists I had ever met. This was prior to our acquisition of Arthur Jaffe. How remarkable was Gilbert's transition to a hands-on biological science, and how successful as well.

By the spring of 1958, I had put together a respectable thesis, and had been awarded an NSF postdoctoral fellowship. I decided to spend a year at the Lebedev Institute in Moscow under the direction of Igor Tamm. This was easier said than done. My thesis exam took place in July at the University of Wisconsin. The committee consisted of Schwinger, Martin, and Frank Yang. The examination consisted partly of a debate between Schwinger and Yang. I had tacitly assumed that the electron-neutrino and muon-neutrino were distinct particles, for three reasons: From the point of view of a gauge theory, the universality of weak interactions demanded separate neutrinos. Secondly, Gary Feinberg, my old high-school buddy who was then and now at Columbia, had shown that the intermediate vector meson hypotheses, in a one-neutrino theory, led to an unacceptable rate for $\mu \rightarrow e\gamma$. Most importantly, the two-neutrino hypotheses was an integral part of Schwinger's teaching. After Schwinger had convinced Yang that my assumption was physically meaningful and predictive, I was excused so that the committee could decide my fate. Three years later, at Brookhaven National Laboratory, Lederman, Schwartz and Steinberger were to prove the truth of Schwinger's two-neutrino scheme.

Since my Russian visa had not yet come, I accepted an invitation along with Danny Kleitman to visit the Niels Bohr Institute for the interim. The interim turned into a very productive two-year stint at Copenhagen and at the European research consortium CERN. The Russian visa never did come, despite the frequent assurances of several Russian consuls that it would appear "tomorrow."

I shall not dwell upon my seven-and-a-half year period of exile from Harvard. Let me jump to January 1966, when I moved from Berkeley to Harvard to continue my fruitful collaborations with Sidney Coleman, my first, best, and somewhat unofficial student. He and I had already successfully explored the electromagnetic properties of baryons in Gell-Mann's unitary summetry scheme. We had also just missed the discovery of the Gell-Mann–Okubo mass formula and of the Cabibbo current. The period 1966–1970 was terribly dull. Although Steve Weinberg, in 1967, brilliantly synthesized the Copenhagen $SU(2) \times U(1)$ electroweak model with the Higgs-Kibble mechanism of spontaneous symmetry breaking, no one seemed to notice or care.

In 1970, we acquired two spectacular postdoctoral fellows, John Iliopoulos and Luciano Maiani. Together, we discovered the GIM mechanism (for Glashow-Iliopoulos-Maiani), an important ingredient of today's electroweak theory. However, our prediction of the existence of a new kind of matter, charm, was not generally accepted. At this point, I took a leave of absence to become a visiting professor at the University of Aix-Marseille, along with John and my student Andrew Yao, now a well-known computer scientist. We were able to remove some, but not all, of the infinities plaguing electroweak models. It was at Marseille that Tini Veltman informed us of the signal accomplishment of his young student, Gerard 't Hooft: the proof of the renormalizability of the spontaneously broken electroweak theory. Things were beginning to happen quickly, so I hurried back to Harvard. In September 1971, I met my wife-to-be as we were shucking corn for a dinner party at the Kleitman's home. Joan, and Danny's wife Sharon, are two of the four notorious Alexander sisters. Meanwhile, experimenters worldwide were racing to discover the predicted neutral currents, and Harvard acquired five very promising young particle theorists: Howard Georgi, Alvaro de Rújula, Helen Quinn, Tom Appelquist, and the precocious graduate student, H. David Politzer.

By 1973, neutral currents were discovered by the CERN group, and were rapidly confirmed at Fermilab by a group led by our very own Carlo Rubbia. Moreover, Carlo's group produced convincing evidence for the existence of a new form of matter, looking tantalizingly like our predicted charmed particles. The situation at the time was reviewed in the form of a play presented in the Jefferson Laboratory on December 3, 1973. Alvaro De Rújula was moderator (an experimentalist), Helen Quinn was speaker (a conservative theorist), Howard Georgi was a talking computer, and I played the role of a model builder.

Roy Weinstein, later the Dean of Science at Houston, invited me to speak at the annual convocation of Experimental Meson Spectroscopists at Northeastern in April 1974. My talk was titled, "Charm: An Invention Awaits Discovery." I offered to eat my hat if charmed particles were not found before the next such meeting. Moreover, the discovery would be made, I claimed, either in neutrino physics or in e^+e^- physics. Sure enough, the J/ψ particle was discovered in the November revolution in 1974, simultaneously at Stanford and at Brookhaven. The Harvard theory group was convinced that charm had been discovered. Indeed, Appelquist and Politzer had essentially predicted the appearance of a charmonium resonance in e^+e^- annihilation. However, the rest of the world was not convinced. As far as I can tell, only Sakharov's group in the Soviet Union, John Iliopoulos in France, and the guys at Cornell shared our confidence and enthusiasm. Not even the observation of a neutrino-produced charmed baryon by Nick Samios in March 1975 convinced the doubters. In fact, in early 1976 a group at the Stanford Linear Accelerator Center (SLAC) published a paper describing their failure to find charm. Fortunately, I was able to convince Gerson Goldhaber to go back and have a second look. He took my advice and by the spring of 1976 charmed particles were finally found in great numbers at SPEAR with properties that were expected of them. At the 1976 meeting of the Conference on Experimental Meson Spectroscopy, I was not to eat my hat. Rather, small and malodorous candy hats were distributed to the participants.

A theory of strong interactions, quantum chromodynamics, had appeared on the scene. It, too, was a non-Abelian gauge theory, but one acting on the hidden variable called color. The theory had many roots and no single discoverer. Gell-Mann's quarks were, of course, essential. So were the observations of Dalitz and Morpurgo on the success of the quark model under the pretense that quarks behaved as bosons. Nambu and Greenberg suggested the notion of a hidden color degree of freedom. One of the crucial steps was the discovery of the asymptotic freedom of non-Abelian gauge theories. This was the key to the understanding of the success of the quark patron model of deep inelastic lepton scattering, and of the narrow observed width of the J/ψ. Asymptotic freedom was discovered by David Politzer at Harvard, and simultaneously by a group at Princeton.

Quantum chromodynamics and the electroweak theory comprise what is now known as the standard theory of elementary-particle physics. It appears to offer, in terms of 17 arbitrary parameters, a complete and correct description of particle phenomenology. There are no loose ends—no observed phenomena that are incompatible with the theory.

At least there were none until March of 1984. CERN is the world's center of high-energy physics for the simple reason that it has made available, for more than a decade, the world's highest energies. The payoff came in 1983 with the discovery at CERN, by Carlo Rubbia and his gang of 137, of the long-sought weak intermediaries $W^{\pm}$ and Z^0. These particles, too, seem to behave just as theory says they should. However, the latest CERN data alleges to reveal an anomaly. At last, there is something that appears inexplicable in terms of the standard model. Surprises, which have been so much a traditional part of the particle game, may come again. Let's hope so.

Author's note: Alas, it was not to be. Better data and more careful interpretation reveal no discrepancy with theory. The standard model reigns supreme. We shall soon see what larger accelerators, now beginning to operate, will have to say.

In 1927 one of my Harvard predecessors, Percy Bridgeman, wrote: "Whatever may be one's opinion as to the simplicity of either the laws or the material structure of nature, there can be no question that the possessors of some such conviction have a real advantage in the race for physical discovery. Doubtless there are many simple connections still to be discovered, and he who has a strong conviction of the existence of these connections is much more likely to find them than he who is not at all sure they are there."

Internal Exile in California

I am, to be sure, a born and bred New Yorker. However, for almost precisely 30 years, but for a brief hiatus, I have lived in Massachusetts and have become, at last, a Bostonian, even to the extent of having once participated in the revels at the notorious Parkman House during the reign of the great Kevin White. More precisely, I am a citizen of Boston's proud and defiant neighbor, Brookline.

Sadly, for five and a half years, I was a subject of American repression, forced to live the hard life of a bachelor in internal exile at the far reaches of the nation, California. I was exposed to starvation in Pasadena (whose restaurants are firmly closed on Thanksgiving Day), to earthquakes in Palo Alto, and to the Filthy Speech Movement in Berkeley. Living from hand to mouth in the land of the setting sun and the silicon goddess was precarious, and I leapt at the opportunity to escape to the slush in January of 1966, for a mere 30 percent cut in salary. All this is history, but even my dark cloud has a silver lining. Few have survived for so long in such a strange land, and have returned practically unscathed. I have become an expert in the ways of the West, a veritable Margaret Mead of California, the first exobiologist with on-the-job training.

Many great discoveries were made in California before my time: artificial elements, cyclotrons, gold mines, and the Pacific Ocean are examples. Let me turn to more recent events in the history of my own discipline, elementary-particle physics.

Many Californians are of Basque ancestry, having been brought to California as sheepherders. Richard Feynman, not himself a Basque sheepherder, hailed from Far Rockaway, which is not all that far from Sheepshead Bay. For this reason, Feynman was the first person to compute the Lamb shift, but he got the wrong answer. The correct result was obtained by my thesis advisor, Julian Schwinger, who only subsequently moved to California, not, at last report, to herd sheep. Feynman's diagrams, the key to his computational skills, have become a mainstay of the decorative arts. They are a prime example of what we physicists call "spinoff"—serendipitous contributions to high technology which are the inevitable accompaniment to abstract thought, the ultimate dollars and cents justification of the leisure of the theory class, the true secret of American know-how. Teflon, Tang, and artificial soda water are other convincing examples of spinoff, gifts of pure science to impure sinners. Who would deny us a mere $5 billion to build the Superconducting Super Collider and continue our grand tradition?*

Donald Glaser is another California achiever. Long ago, while pondering the curious behavior of his *N*th beer, he concocted the bubble chamber, a notorious and dangerous instrument which led to the discovery of hundreds of new and unwanted elementary particles. This produced a profound dilemma: Were all these particles equally, and hence not very, elementary? This was the view of Geoffrey Chew, leader of the renowned school of Maximal Analyticity and defender of Nuclear Democracy. The only alternative was quarks, which not even Murray Gell-Mann could correctly pronounce all of the time. In the end, quarks emerged victorious and even managed to purge themselves of their more radical fringe, the Quark Liberation Front, and to implement a successful affirmative action program with respect to color. It sounds like a happy story, but remember this: Nuclear Democracy is dead, as dead as the nuclear family, and the population explosion of elementary particles continues unabated. All this because of Dan's bubble chamber. The moral: Never do today what might never get done tomorrow.

Understandably, Luis Alvarez soon tired of the many new particles being discovered in his bubble chambers. "Seen one, you've seen them all," he said. Thanks to his seemingly Spanish surname, he set off to Egypt to invent the CAP scanner: computer assisted pyramidology. Heroically, utilizing the most powerful cosmic rays, he searched for secret chambers in the great Pyramid of Cheops, and dreamed of the untold wealth of the Pharoahs that might be his. The search was in vain, and Alvarez returned to California, a broken and embittered shell. Together with his son, he conspired to reveal the secret of how the dinosaurs died, and he predicted, beyond a reasonable shadow of doubt, that the same

* *Author's note:* The cost of the SSC has since risen to $10 billion.

would happen to us. In precisely 13 million years (and this, incidentally, explains the mysterious but well-founded fear of the number 13, or tristadekaphobia, to physicians and songwriters); in precisely 13 million years the Sun's dread, but invisible, spouse Nemesis will return to reap its destructive harvest, and to end the world as we know it. Foolish astronomers, even now, are searching the skies for Nemesis. When they find it, perhaps they will foretell the precise day, hour, and minute of our species' inevitable demise. Because Alvarez had to tell it as it is, all human hopes and aspirations for the future have been crushed.

Kendall, Friedman, and Taylor, laboring at the Stanford Linear Accelarator Center, a small atom smasher merely two miles long, patiently repeated the experiments Ernest Rutherford performed 56 years earlier, but at a somewhat higher energy. Just as their predecessor discovered the atomic nucleus, they found evidence for the existence of pointlike particles within the very proton, pointlike particles they pronounced partons. They were, of course, mistaken. Their pointlike particles were none other than the quarks that Gell-Mann had prophesied but abandoned years before. Despite their error, Kendall and Friedman have escaped exile, but sad to say, Dick Taylor still languishes in the golden state.*

We come, inevitably, to the great Goldhaber clan, America's answer to the Swiss Bernoullis, and to Gerson of California in particular. It was the cold winter of '76, and most citizens were scrimping and saving to buy fireworks with which to celebrate our Bicentennial. The Goldhabers were far too busy to concern themselves with the decisive battle being fought between the charmed quark and the Wicked Witch of the West. The Witch had gone so far as to deny charm's birthright by official proclamation in *Physical Review Letters*. Gerson, though he was at Berkeley, was our most valued double agent. We arranged a last clandestine meeting in April in Wisconsin. The Experimental Meson Spectroscopy Conference was scheduled for June. If charm were not found by then, they would force me to ingest my chapeau. Time was short. Our desperate plan succeeded, and charm was revealed on schedule. The Wicked Witch of the West had been not merely conquered but converted, and she designated Gerson as California Scientist of the Year. He was also awarded the highest honor bestowed by the University of California—an orange circle parking sticker.

* *Author's note:* Kendall, Friedman, and Taylor shared the 1990 Nobel Prize in Physics for their discovery of quarks.

All these great discoveries notwithstanding, it is not surprising that some experiments, especially in California, do not stand quite so well. I have written a small poem, somewhat distasteful, that celebrates some of these results, to wit, the hunt for fractional or magnetic charges and the discoveries of remote viewing and of anomalons. The poem ends upon a more speculative note. But first, a little background:

James Clerk Maxwell did not believe in the existence of fundamental particles carrying charge: neither electric nor magnetic. Nevertheless, by recognizing the self-evident symmetry between electricity and magnetism, as in the old saw "Why is a cat rubbed against a window like a compass?" he set us upon the road leading inexorably to the great grand unification of today and to microwave ovens. In 1896, the renowned Irish physicist George Johnstown Stoney, flushed by his rediscovery of the wheel which brought the Industrial Revolution to Ireland, named the elementary carrier of electric charge after his young lady friend Amber. Once named, the electron was observed in the laboratory only one year later. Maxwell would have been outraged, had he not been dead for almost two decades.

This brings us to Paul Adrien Maurice Dirac, once unanimously acclaimed as the World's Greatest Living Physicist, whose famous textbook on quantum mechanics was rejected by the prestigious Cambridge University Press. In retaliation for this slight, Dirac sent his book to the rival Oxford Press, and transferred his allegiance to Florida State University in Tallahasse. In between, he realized that electromagnetic symmetry demanded that Maxwell could neither be half-right nor half-wrong. The existence of the electron demanded that there be magnetic charges as well.

All of Dirac's predictions have come true. For this reason, experimenters discover magnetic monopoles every few years, only to retract them later for technical reasons. My colleague, Ed Purcell, is the only experimental physicist who has looked for magnetic monopoles but never found one. For this, he was awarded the coveted Nobel Prize in physics.

Quarks have fractional electric charges, but you are not supposed to see them. Nonetheless, many experimenters search for fractional charges. They hope to incite yet another revolution on physics. Observable fractional charges and Dirac magnetic monopoles bear a curious relationship to one another: either one of them could in principle exist. One but not both. Thus it is remarkable that these two mutually exclusive entities have apparently both been reported by scientists working at the same laboratory in California.

Of the alleged observation of fractional charges, perhaps the less said the better: Experiments are a decade old and have neither been confirmed nor retracted. The one magnetic monopole candidate appeared on St. Valentine's Day, 1982. A year later, I sent the following telegram to the investigator, Blas Cabrera:

Roses are red; violets are blue.
The time is now
For monopole two.

I have not yet received an answer.

Certain theorists, known as neonatal cosmologists, have clarified the subject by establishing two alternative scenarios. (A scenario, incidentally, is what you call a theory when you've got more than one.) In one scenario, the universe was created chock-full of magnetic monopoles. There were so many that billions of years ago the universe ceased to exist. In the second, perhaps more plausible, scenario, there is just one magnetic monopole in the entire observable universe—perhaps it passed through Cabrera's apparatus on St. Valentine's Day in 1982.

A third remarkable result emerged in the environs of Palo Alto, at SRI which was known as the Stanford Research Institute, until its goals diverged from those of the university. In high school, I was a fan of J.B. Rhine and of his demonstrations of extrasensory perception at Duke University. While his experiments were inconclusive, those of Putoff and Targ at SRI were convincing enough to be published in the *Proceedings of the IEEE*. The human talent of remote viewing permits the uninformed individual placed in a sealed room to describe the observations of a distant colleague. A program of experiments designed to confirm this effect was proposed by my colleague Paul Horowitz in collaboration with the Amazing Randy. It was deemed to be unnecessary by the SRI scientists.

Finally, I come to the exotic discipline of nuclear physics, a branch of physics that is not actively pursued at Harvard University. Few of my colleagues realized that there are precisely $3 \times 7 \times 19$ known species of atomic nuclei that live for at least one year. Even at MIT, interest in nuclear physics appears to be waning, for they did not even submit an expensive proposal to the latest nuclear accelerator sweepstakes. One reason for this deplorable lack of interest in a fundamental science is the fact that the latest half dozen synthetic chemical elements, quite unlike plutonium or curium, have not even been given names, only numbers. "Do not fold, spindle, or mutilate me!" cry their neglected nuclei. It is in this context that the nuclear scientists of California have come to the fore to fill a much needed gap in the literature. They have discovered a new effect, indeed, a new form of nuclear matter that has no conceivable rational explanation: the *anomalon.* Once again, like the fractional charge, the monopole candidate, and remote viewing, the anomalon is a peculiarly Californian experience. Like so many allegedly great California wines, it does not travel well.

The last experiment I shall discuss is the reported observation of the zeta particle that was tentatively announced at the International Conference on High Energy Physics at Leipzig in the summer of 1984. The results emerged from the Crystal Ball detector operating at the German electron synchrotron center in Hamburg. The Crystal Ball detector was designed and created in California by Californians. The director of the laboratory and I agree that the discovery of the zeta particle would be, if correct, the most significant event in particle physics in the past decade.*

Now, a poem:

To Abalone Unbound

What with chromodynamics and electroweak too,
Our standardized model should please even you.
Tho' once you did say that of charm there was none,
It took courage to switch, as to say, earth moves, not sun.
Yet your state of the union penultimate large
Is the last known haunt of the fractional charge,
And as you surf in the hot tub with sour-dough roll,
Please ponder the passing of your sole monopole.
Your olympics were fun, you should bring them all back—
For transsexual tennis or anomalon track,
But Hollywood movies remain sinfully crude,
Whether seen on the telly or remotely viewed.
Now fasten your sunbelts, for you've done it once more.
You said it in Leipzig of the thing we adore,
That you've built an incredible crystalline sphere
Whose German attendants spread trembling and fear
Of the death of our theory by particle zeta,
Which I'll bet is not there, say your articles, later.

* *Author's note:* Soon after this essay was written, the alleged discovery of the zeta particle was retracted.

The Mysteries of Matter

WITH PETER COSTA

The following is an edited transcript of a conversation between Glashow and Peter Costa, director of Harvard's Office of News and Public Affairs.

Before we talk about physics, I'd like to talk to you about some of the personal things you wrote about in your recent autobiographical book, Interactions. *In the book, you discuss your transformation from a "nerd"—I believe that is what you called yourself—at the Bronx Science High School in New York to a Nobel Prize winner. Throughout the book, you credit your father, a plumber and a craftsman, for instilling in you a lifelong curiosity about how things work. How do you manage to keep the fires of curiosity burning after all these years of scientific investigation?*

GLASHOW: That is a difficult question. You mention "nerds." Of course, "nerds" are an endangered species today, and I'm a great defender of them. I go hunting for "nerds" because they make promising graduate students, eventually.

Now, how could I still be interested? I got into physics because I found it so much more interesting than anything else. I was not very good at sports . . . so I picked physics as something that was a lot of fun. And it still is. It's very exciting and certainly addictive.

I think I'm not as smart as I used to be, but I'm certainly at least as much interested in these questions. They are always the same questions: What is it all for? What is it all made of? How does it all work? They still fascinate me.

Did you look at, say, a radio when you were growing up, and want to take it apart?

GLASHOW: Oh, yes. I took clocks apart, I took radios apart. But I could never get them back together again. There would always be something left over that didn't fit and it wouldn't work. So that's why I'm not in experimental physics. I had the most marvelous electric train set, vintage 1930s, which I managed to take apart and I could never get back together again. My experimental life is full of tragedies.

So you kept taking things apart until you reached the subatomic level, so to speak.

GLASHOW: So to speak.

In your book, you explain the intricacies of subatomic particles, bosons, and quarks, but you also tantalize us with a story about your dad and how he survived falling into a vat of molten lead. Tell us more about that.

GLASHOW: My father came to the U.S. from Russia in 1905. He first took all kinds of laboring and construction jobs. At one point, he was building a house, or putting plumbing into one. They used molten lead on the joints and he simply fell into the tank of lead. He explained to me that he was protected by a tiny layer of air, so he didn't get seriously burnt.

I was impressed by the physics and by the fortuitous accident in which nothing serious happened.

Your father came through Ellis Island during the wave of immigration. What kind of transformation went on there? Some came out with names that were not theirs.

GLASHOW: Our name in Russia was Glukhovsky. Neither the immigration officer, nor my father thought that that name would do in America. So they had an amicable discussion and they puzzled things out—in a totally friendly fashion, so my father said—and came up with this bizarre name, Glashow. Now there are hundreds of Glashows.

Tell us about the driving force to succeed that often comes from being of immigrant parents. You have two brothers, one of whom is a doctor, and the other a dentist, both very successful. Did the expectations concerning what is success and what is not affect you as a boy?

GLASHOW: My parents were very concerned about how well I did in school as I am with my kids. But I don't think it was the parental influence as much as the environment of our middle-class, half-Jewish, half-

Irish Manhattan neighborhood. There was a strong desire to succeed, to pull out of our mean existence. The other children too have been very successful—many of them doctors, dentists, lawyers, and scientists. The whole circumstance was one that emphasized learning.

Just as there is, today, a strong tendency not to stand out, not to accomplish anything, and to not be a "nerd," at that time it was not so much the desire to be a "nerd," but the desire to triumph in school.

Education was the vehicle for upward mobility.

GLASHOW: Absolutely.

You've written that there is no intellectual pursuit more challenging than physics. What is it about physics that makes it so difficult?

GLASHOW: No, it's not a difficult science. I think modern biology is much more difficult: you have to remember the names of all kinds of god-awful chemicals. You don't need that in physics. The thing about the kind of physics I do—which is fundamental physics—is that we don't know the rules. It's a contest, a game. It's a kind of gambling game, where you put your money where your mouth is concerning what you think is the way nature works. Then, if you're lucky, you get proven right. There is no greater feeling than winning this bet with nature.

Once upon a time, my buddies and I figured out that there had to exist a fourth kind of quark, and 10 years later it was found. That feels pretty good. It's the kind of feeling that I've had a few times in my life. It's a challenge, a game. It's like going to a magic show and figuring out how the tricks are done. It's like reading a detective story and trying to anticipate the ending. It's all of these things. It's a very human activity, except that it is focused on one simple question: how does it all work?

But your detective story presupposes you possess the language of very advanced mathematics?

GLASHOW: It presupposes the knowledge of not so advanced mathematics. I can't understand physics, unless I understand mathematics; I can't understand chemistry, unless I understand physics. This is a very important fact because, as you know, Harvard is divided into three parts: its sociologists, its humanoids, and its scientists.

At Harvard, you get to talk to people in other disciplines. But scientists must enter the others' turf because other scholars deal with horizontal disciplines where, in a sense, they possess shallow knowledge of a wide variety of things. But ours is a vertical discipline where everything depends upon everything else. They never know science. We know a little bit about literary criticism, or about history, or about ecology or about sociology. We're always compelled to discuss things or argue things or enjoy conversations on their turf, never on ours, because they lack these skills.

It's true we live in two societies and in a sense it's not true. There is one society: there are those who are literate and understand things, and there are the rest, who don't. Unfortunately, two-thirds of Harvard misses the best part of human knowledge.

Perhaps now is a good time to mention your plea for scientific literacy among students.

GLASHOW: It's a mess. But illiteracy is not just in science, it's not only in math, as you know, but most are illiterate about history, too. I would venture to say that less than half the population of the U.S. knows the relative time order of the American Revolution and the Russian Revolution. Very few people can identify Rome on a map of the world.

It's worse in science. Some try to improve science education. The National Science Teachers Association is very ambitiously attempting to unify all of science education. All of science can fit together into a unified meaningful whole, instead of teaching bits and pieces, here and there.

Why are there so many talented mathematicians, originally trained in physics, but comparatively fewer physicists who come from mathematics?

GLASHOW: Generally, the tendency toward abstract mathematics is irreversible. There are physicists who move into more abstract mathematics circles. Einstein is a good example. Heisenberg was another. They began with down-to-earth questions. They do quite well at solving those problems and then move on to more abstract ones. It's very rare to go the other way. Physicists often move in another direction. There has been a very large emigration of physicists into biology. Walter Gilbert is an example, and there are a dozen others, of those who began as physicists and moved into biology.

Let's turn to the need for university professors to teach as well as to do research. You teach a "Core" course at Harvard in physics for undergraduates. I studied freshman physics with a friend of yours, Leon Lederman. I remember one of his lectures about the conservationist energy which involved a 200-pound brass sphere suspended from a wire from the ceiling in the lecture hall. Lederman walked to the side of the lecture hall and very slowly and carefully placed his back and the back of his head against the wall. Then he had an assistant move the sphere near him. He grabbed the sphere between his hands and brought it right to the tip of his nose and very gently released the sphere and let it slip through his hands. The sphere, of course, swung across the room in a great arc to the far wall, swung back and stopped short, in what seemed just microns, in front of his nose. We all gasped and applauded madly. Lederman was always doing things like that.

GLASHOW: I often wondered where Leon got that funny looking nose . . .

Making physics fun for us.

GLASHOW: That was a demonstration I witnessed in class in Cornell in 1950. I also perform it from time to time.

Sometimes we ask a graduate student to lie down on a bed of nails and smash a large cinder block over his chest. Once I hid a plastic container of ketchup on the student's chest and he got up, covered with "blood." But it's hard to get graduate students willing to sacrifice their shirts.

You have said that it is important for a teacher to be a researcher to enliven theory with practice.

GLASHOW: I don't think there are clear boundaries between teaching, research, and learning. As we teach, we learn, and as we do research, we present the research in teaching. As students respond, we learn how better to do the research. It's all folded together. That's what major American universities are all about. They mix these activities, and so they should.

In your "Core" course, what would you like your students to come away with?

GLASHOW: I would like them to be excited about, and to understand a little bit of, how it is that people have been able to understand (to the extent they do) what matter is made of, what goes on in a brick as you get ten times closer, as you blow it up in size by a factor of 10, then another factor of 10, then another and another and another; to understand exactly how it's put together.

And, conversely, if you imagine lots of bricks, the whole world, the solar system, the galaxy, the universe itself, to reconcile the birth and evolution of the whole of the universe with the properties of matter on a very elementary scale.

At the moment that's what it is all about: the physics of the very large and the physics of the very small have come together. The snake is in the process of eating its tail, and it's very exciting to be around at this time. Everything is finally being put together.

Could you tell us something about what is going on in a subatomic level—what the weak force is, are there things smaller than quarks?

GLASHOW: That is a lovely question. Every time I get asked the question I wonder "Does he understand air pressure? Does he understand why Boyle's law is true? Why there is a spring in the air?" Many times the answer is no, and if it's no, it's very hard to go further and explain what the weak force is, what the strong force is.

Once upon a time (about the 1930s) we realized that all of the many different displays of force and motion are reduceable to just four forces which we call, briefly, the strong and the weak nuclear force, gravity, and electromagnetism. Actually, gravity and electromagnetism are about all you need for most anything.

Plumbing, my father's specialty, depended on gravity because water goes down, and it depended on electromagnetism because it explains how you wipe a joint and why copper is and what it is, and everything else about everything you see, feel, smell, touch, or do. Ultimately, you do need the other two forces. The strong and weak nuclear forces. They have to do with the atomic nucleus and radioactivity. But they're important because, if you didn't have a nucleus, you wouldn't have an atom, and we wouldn't have anything else.

So the strong and weak forces have to do with the atomic nucleus. What we who won the Nobel Prize some years ago (and a number of other people who didn't) realized is that two of these four forces are really different aspects, different avatars, of the same underlying equation or system; mainly, that the weak force and electromagnetism are really one.

So, in a sense, we've reduced the number—and only in a sense—of forces from four to three, suggesting a further reduction from three to two and, of course, some ultimate unified dream of Einstein—one. But we ain't there yet.

So much of experimental and theoretical work in physics depends on giant pharaonic-sized machines. Will these mega-machines help you in your work as a theoretician?

GLASHOW: We depend on experimental information, and all these wonderful things we know about nature and the universe, depend upon telescopes, X-ray observatories, accelerators, and other devices. So I am tremendously excited about the new accelerators that are to come into operation in Europe. I'm even more excited about the Waxahachie initiative, the Gippertron—Ronald Reagan's accelerator in Texas.

I didn't know it was the Gippertron.

GLASHOW: Not officially. It's the Ronald Reagan Center for Particle Physics. If it's funded . . . .

Some scientists think that if we take the billions of dollars that might go into that machine and put the funds into something else . . .

GLASHOW: Well, I don't see why they want to take my money away. I agree that science is underfunded and should be doubled. That's just about what we're asking for in particle physics. We can build the Superconducting Super Collider (SSC) with double the present budget for high-energy physics. As far as the other scientists go (who are doing wonderful work), their budgets should be doubled, too. But I didn't want to build my machine by taking away their funds. I doubt whether they want to do their work by taking away our money.

Is it more difficult to persuade Congress to support this project because we're not going to get, say, Teflon out of it?

GLASHOW: Teflon was pretty good. That came from the atomic bomb project. Did you know that? In separating uranium, which was something necessary to do in order to get U-235 to make bombs—which is what was wanted in those days—they performed experiments with heavy gases, uranium gases. You make a gas out of uranium by combining it with fluorine. One of the things that came out of the Manhattan Project was, in fact, Teflon. I'm sure that all sorts of wonderful technology will flow out of building the SSC: superconducting energy storage, mass transit, tunnel development, all kinds of technologies at the cutting edge.

I detect that you may be a little bitter about having to share your Nobel Prize, unlike in Madame Curie's day. Now there seem to be two and three people who share the prize.

GLASHOW: It's true the person who gets the whole prize gets three times more money than a person who doesn't and has to share it with a couple of other people. In fact, you can even get a quarter of the money.

But the money is not the issue with the Nobel Prize. It's just a queen for a day or a king for a day. It's really quite wonderful, a great honor. It's nice to be honored. But I'm very happy that I shared my prize with two good friends, Steven Weinberg and Abdus Salam, who were, and remain, my very good friends.

How are the Europeans doing in theoretical physics compared to us?

GLASHOW: It's hard to pin that down exactly. At the moment, the best experimental facilities are in Europe and that's very inspirational to European physicists, so they do very well. The Italians do superbly, the French do very, very well in condensed-matter physics. Everybody's doing well. The Russians are marvelous. It's not a competitive game. It's always been the most international sport of all—the pursuit of physics.

. . . Physics was international in the seventeenth century and so it is today. We treasure our international fellowship very much and we rarely try to say, "This was done in America, this was done in Europe, this was done in Russia." It was done by us, working together.

There are very few women in theoretical physics. Is there anything that can be done about it?

GLASHOW: That's changing and yes, there are things you can do about it. The solution is to have women who are interested in physics learn physics at an early enough stage. But there are lots of women and they do very well. Some have been eminently successful. So it's changing. At Harvard we have 25 or 30 percent women in my field. We don't yet have 50 percent.

THE WORLD OF SCIENCE

A Peek at the Universe

We may debate endlessly whether or not dolphins, chimpanzees and other beasts possess the power to reason. They can be taught to do tricks, solve puzzles, and, after a fashion, talk. One thing, however, is certain: Only the human being is capable of wonder. Ours is the only species on earth to observe the splendor of the heavens and to demand to know the reason why. Primitive societies discovered or imagined patients among the stars: the lion, the twins, the crab, and so forth. They sought to give meaning to human experience in these constellations. Even our language today reflects the ancient superstitions: The words "consider," "mercurial," "venereal," "lunacy," "martial," "jovial," and "saturnine" each correspond to one of the seven heavenly bodies, while seemingly everyday English words like "disaster," "influence," "hour," and "contemplate" are also of astrological origin. In our advanced and enlightened world today, there are still those who seek their fortunes in the sky. The persistence of so pernicious a pseudoscience as astrology makes one doubt whether the power to reason is a truly universal human ability.

Early societies discovered that the stars and the Milky Way, the entire firmament containing the 12 principal constellations, seems to rotate rigidly about the earthbound observer. Each star, in the course of an evening, moves along an arc of a circle. The photograph in Figure 1, made with a long exposure, illustrates this elementary truth. To the ancients, this is evidence for a fixed and immutable celestial sphere upon which the stars are attached, and of its daily rotation about the Earth.

The Greeks believed that various forms of matter on Earth could be explained as mixtures of four essential ingredients: fire, water, Earth, and air. There was a fifth fundamental material, the "quintessence," ex-

FIGURE 1. *A long-exposure photograph of star trails around the north celestial pole.*

plained the heavens. The motion of this perfect and heavenly substance could only be in a circle, the perfect figure with no beginning or end. The rotational motion of the celestial sphere was explained, and the Aristotelian vision of the cosmos was to reign for more than a millenium.

Not every heavenly body was accounted for. There were precisely seven more that were evidently not part of the celestial sphere: the Sun, the Moon, and the five planets. (*Five,* not nine; three were yet to be discovered and Earth was a very special place.) The profound significance of these celestial objects to ancient societies is reflected today in the names of the days of the week, given in Table 1 in three representative Western languages.

TABLE 1. Names of the Days of the Week

English	*French*	*Latin*
Sunday	Dimanche	Dies Solis
Monday	Lundi	Dies Lunae
Tuesday	Mardi	Dies Martis

English	*French*	*Latin*
Wednesday	Mercredi	Dies Mercurii
Thursday	Jeudi	Dies Jovis
Friday	Vendredi	Dies Veneris
Saturday	Samedi	Dies Saturni

In Latin, each day of the week corresponds to one of the heavenly bodies, and to its representative deity. In French, this correspondence is maintained, except during the weekend. The English day names are direct translations of the Latin, with the Anglo-Saxon deities Tiw, Woden, Frigg and Thor usurping their Latin equivalents. Curiously enough, Venus (the token woman deity) is replaced by Frigg, a.k.a. Mrs. Woden. The histories of religion, timekeeping and astronomy are intricately woven together.

The seven heavenly bodies move across the sky in a much more complex fashion than do the fixed stars. The Moon travels in circles, but different circles than the stars. This was easily accommodated by imagining the Moon to be attached to a second celestial sphere with its own circular motion. So also for the Sun, embedded upon a third serenely rotating sphere. The word ''planet'' stems from the Greek word for wanderer, and wander they do. Some of the planets even reverse the sense of their motion across the sky from time to time. The simple notion of seven rotating and concentric spheres had to be abandoned.

Still, circular motion was philosophically mandated for the quintessential heavenly bodies. Tricks were devised. Planets were to move in circles, but circles not centered about the stationary Earth: *eccentric* motion. Planets were to move in circles within circles: *epicycles.* Planets were to move in circles, but not at a constant rate as viewed from the center: *equants.*

Using eccentrics, epicycles, and equants, Ptolemy, with a great deal of hard work, formulated his system of heavenly motions in the second century A.D. The Earth was immobile in the very center of the universe. The past and future motions of the planets could be explained. Serious discrepancies between prediction and observation were resolved by small adjustments of the 70 simultaneous and independent motions of the seven bodies. While it was incredibly intricate and contrived, and the Ptolemaic system was understood by only a few, it was generally accepted; it worked, and it put Earth right in the middle, where Earth belongs. After the Dark Ages, when Arab guardians returned the Ptolemaic system to Europe, it soon became part and parcel of Christian dogma.

Nicolaus Copernicus (1473–1543) was a devout scholar and a canon of the Roman Catholic church. A firm believer in the primacy of circular motion, he was disturbed by the complexity and arbitariness of the Ptolemaic theory. To Copernicus ". . . the planetary theories of Ptolemy and most other astronomers, although consistent with the numerical data seemed . . . to present no small difficulty . . . Having become aware of these defects, I often considered whether there could perhaps be found a more reasonable arrangement of circles . . . in which everything would move uniformly about its proper center (Copernicus despised equants), as the rule of absolute motion requires."

Copernicus realized the price that must be paid for a simpler system was to abandon the fixed Earth, to put the Sun in the center of the universe, and to have the rotating Earth move in a circle about it. Far from putting forth such a radical notion himself, he appealed to prior authority, " . . . according to Cicero, Nicetas had thought the Earth moved . . . according to Plutarch certain others had held the same opinion . . . when from this, therefore, I had conceived its possibility, I myself also began to meditate upon the mobility of the Earth. And although it seemed an absurd opinion, yet, because I knew that others before me had been granted the liberty of supposing whatever circles they chose in order to demonstrate the observations concerning the celestial bodies, I considered that I too might well be allowed to try whether sounder demonstrations of the revolutions of the heavenly orbs might be discovered by supposing some motion of the Earth."

The Copernican system was at least as predictive as Ptolemy's and very much simpler. All the planets, Earth included, could be thought to move on concentric spheres. Only a few small epicycles and eccentrics had to be added to give quantitative precision.

Copernicus' actions were completely consistent with his firm religious beliefs. His heliocentric theory was but a reflection of the mind of the Creator, its greater simplicity but a reaffirmation of the principles of Aristotle. Yet, Copernicus made very few converts. The old church and the new were firmly committed to the geocentric view. For is it not written in Scripture. "Then spoke Joshua to the Lord in the day when the Lord delivered up the Amorites before the children of Israel, and he said in the sight of Israel, Sun, stand thou still upon Gideon; and thou Moon in the valley of Ajalon. And the Sun stood still, and the Moon stayed, until the people had avenged themselves." If the Lord could put the Sun temporarily to rest, then surely the Sun must move. Perhaps the Copernican system was useful to the church as a computational tool (the calendar was getting a bit out of whack), but as a statement about the real world, it was heresy. Eventually, the Catholic Church was to put Copernicus' magnum opus on the index of forbidden books as "false and altogether opposed to Holy Scriptures" where it was to remain, banned to practicing Roman Catholics, until 1835.

Johannes Kepler (1571–1630) was a convinced Copernican, whose purpose in life was the perfection of the heliocentric theory. Even more than Copernicus, he believed in the underlying simplicity of physical laws. For example, he asked why there should be exactly six planets, no more, no less. (Earth has become merely another planet in the Copernican view: a degrading heresy, but nonetheless true.) Kepler knew that there exist just five regular geometric solids: tetrahedron, cube, octahedron, dodecahedron, and icosohedron. Place these five solids within a series of six nested spheres, and the radii of the spheres will be in proportion to the solar distances of the six planets. This example of Kepler's laws is pure nonsense of course, but it is a true indication of his boundless imagination.

While the telescope had not yet been invented, the quality of astronomical data was far superior to what was available to Copernicus. This is largely due to the painstaking work of Tycho Brahe at his island observatory off Denmark. Measurement errors had been reduced by more than a power of 10. Kepler set himself the problem of fitting the new and precise data on Mars' orbit to his believed Copernican theory. After four years of study, he realized that the task could not be done—the theory was, quite simply, wrong. The motion of Mars is not a circle, nor even a circle as modified by epicycles, eccentrics, or equants. It is an ellipse. With this discovery, Kepler's faith in the fundamental simplicity of natural law was vindicated. The motions of each of the six planets could be fitted to a simple ellipse. No Ptolemaic tinkering was needed: eccentrics and epicycles joined equants in the great garbage heap of discarded physical theories.

Planets move in ellipses with the Sun at one focus. This is the first of Kepler's three great empirical rules, replacing the Aristotelian principle of circular motion. Planets, theologically speaking, were permitted to move in such imperfect orbits, since Kepler believed that they themselves were imperfect material bodies like Earth.

The second law replaces the notion of uniform motion: the line between the Sun and a planet sweeps out equal areas in equal times. The third law is the stunningly simple statement that the square of the period of a planetary orbit (its year) varies with the cube of the mean radius of its orbit. This law has no Aristotelian precedent, since it relates the motions of different planets to one another.

Another way of phrasing the third law is to say that the mean speed of a planet varies with the reciprocal square root of its distance from the sun. To illustrate this law, let us provide in Figure 2 a graph of planetary speeds versus their distance from the Sun. See that the nine points lie on a smooth curve corresponding to $1/\sqrt{R}$. The planet Mercury, being 1 percent of the distance from the Sun as Pluto, travels in its orbit 10 times more rapidly.

None of Kepler's three great laws was deduced from a consistent theoretical framework—this would be accomplished a century later by Isaac Newton. Kepler searched for and found his relationships by careful analysis and inspired guesswork. He was perhaps the most brilliant theoretical astronomer of any age, just as Tycho Brahe may have been the best observational astronomer ever. Kepler and Brahe preceded the institution of the Nobel Prize by three centuries (and there isn't even a prize for astronomy.)

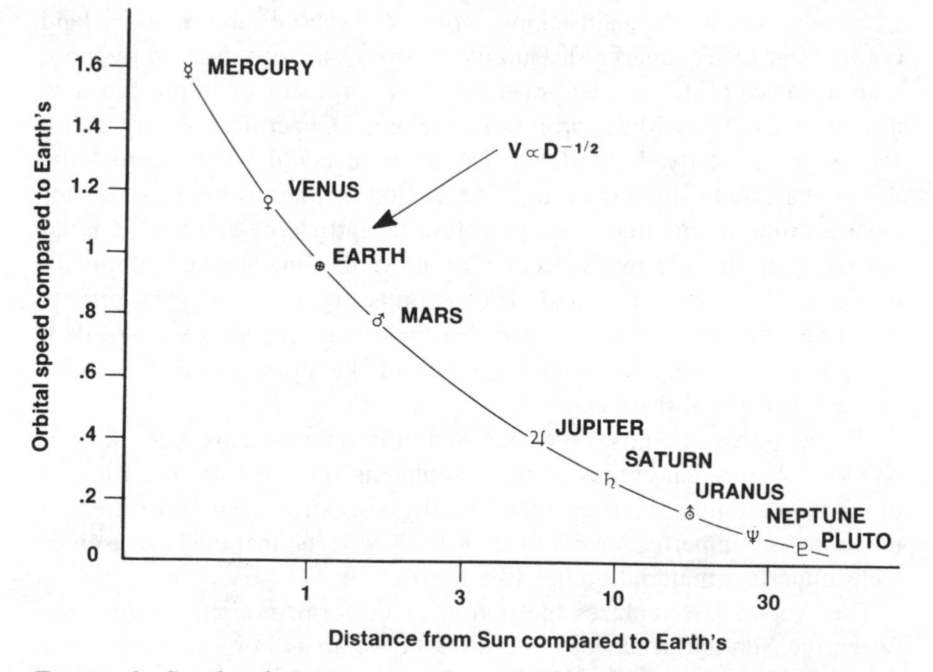

FIGURE 2. *Graphs of planetary speeds versus their distance from the Sun.*

Supernovae are titanic stellar explosions, so great they may outshine the galaxy in which they appear. Rarely are there supernovae in our galaxy. One of the most spectacular took place in the year 1054. Chinese observations record that the star was so bright that it was visible all through the day, and could compete with the Moon at night. No European report of the event has been discovered. How great are the fear and blindness that a totalitarian faith can engender!

The most recent supernovae in our galaxy took place in 1572 and 1604. They are known as Tycho's supernova and Kepler's supernova. No astronomer since has earned such an award.

Galileo was born in the year Michelangelo died (1564) and he died in the year of Newton's birth (1642). Like Kepler, Galileo was a believer in Copernicus. Not nearly as mathematically inclined as Kepler, Galileo was never to accept his colleague's notion of ellipses. In 1609, having heard about the Dutch invention of the telescope, Galileo constructed one himself. He immediately realized its potential as an instrument of war: it was to provide a "distant early warning" of approaching enemy ships. The military-industrial complex of the Republic of Venice paid handsomely for the device.

Galileo's first telescopic observations confirmed his Copernican prejudices. Three examples: The Milky Way to the naked eye is just that—a diffuse, continuous, milky band across the sky. Through the telescope, it is revealed as untold thousands of individual stars. (Today we know that our galaxy, the Milky Way, contains about 10^{11} stars.) But, faith decreed that the universe had been fashioned solely for the joys and sorrows of mankind. For what purpose are these invisible stars which can only be seen by means of magic devices? (The telescope, being of evident military value, was never accused of being an accursed instrument of the devil.) Of our nearest celestial neighbor, Galileo wrote "the surface of the Moon is not smooth, uniform, and precisely spherical as a great number of philosophers believe it (and the other heavenly bodies) to be, but is uneven, rough, and full of cavities and prominences, being not unlike the face of the Earth, relieved by chains of mountains and deep valleys." Galileo saw blemishes on the surface of the Moon and Sun. Heavenly bodies were not so perfect as faith decreed.

The most startling of Galileo's discoveries was that the four principal satellites of Jupiter: Io, Europa, Ganymede, and Callisto revolved about the planet. Their periods were found to obey Kepler's third law—an astonishing success of his model in an unanticipated new domain. No rational person, viewing Jupiter and its moons, could any longer doubt the heliocentric theory.

Galileo's work was not universally accepted. Tricks could be played with lenses. Indeed, they must have been since Galileo's conclusions were metaphysically impossible:

> There are seven windows in the head, two nostrils, two ears, two eyes, and a mouth; so in the heavens there are two favorable stars, two unpropitious, two luminaries, and Mercury alone, undecided and indifferent. From which and many similar other phenomena of Nature such as the seven metals, etc., which it were tedious to enumerate, we gather that the number of planets is necessarily seven . . . Besides, the Jews, and other ancient nations, as well as modern Europeans, have adopted the division of the week into seven days, and have named them from the seven planets: now if we increase the number of planets, this whole system falls to the ground . . . Moreover, the satellites are invisible to the naked eye, and therefore can have no influence on the Earth and therefore would be useless and therefore do not exist.

So argued the Florentine astronomer Francesco Sizzi in 1611.

Useless and invisible to the naked eye they may be, but Galileo's "Medicean Stars" (named after his patron) were there to be seen by anyone with access to a telescope. In the mid-seventeenth century, five moons of Saturn were added to the celestial bestiary. They too, satisfied Kepler's laws. Why did these simple laws work so well? The key to the puzzle was the realization that the natural state of motion of an undisturbed object, whether on earth or in the heavens, is a straight line. Secondly, it was necessary to realize that the force which makes bodies fall to earth extends into the heavens and is responsible for the motions of planets. Thirdly, powerful new analytical techniques were needed to deduce the nature of a planetary orbit from the assumed form of the force.

Isaac Newton possessed the brilliant mathematical talent needed to solve a problem that had baffled all before. Moreover, he was utterly convinced of the universality of physical laws. Celestial mechanics and terrestrial mechanics were, to Newton, one and the same discipline. Newton's hypothesis of a universal gravitational force, that two bodies attract one another with a force proportional to the mass of each body and inversely proportional to the square of the distance between them, enabled him to deduce Kepler's three laws. Much more than this, Newton showed that the very same force explained the motions of planets about the Sun, satellites about their primaries, terrestrial falling bodies, the movements of the tides, and even the trajectories of comets. The mystery of the motions of celestial bodies had been solved with purely terrestrial concepts. Or, rather, many mysteries had been reduced to one simply stated mystery: What is gravity? Published in 1687, Newton's work remains today as the most momentous scientific revelation of all time.

The seventh planet was discovered by one William Herschel, son of an oboist of the Hanoverian Footguards Band and an accomplished professional musician himself. In his spare time, Herschel built telescopes. Not ordinary telescopes, but the very best telescopes of his time. Herschel set himself the stupendous task of performing a complete and systematic survey of the heavens. In 1781, he discovered a curious new object which be believed to be a comet. Within months, it became clear that Herschel's comet was no comet at all, but a new planet. But the question of the name on the new planet was vexing. Herschel called it "The Georgian," in honor of his patron, King George III, and so it was known in Great Britain. The French insisted on the name Herschel, in honor of its discoverer, and so it was known in France. Swedes and Russians suggested the name "Neptune," but a careful reading of mythology suggested a better alternative. Uranus, god of the sky and husband to Earth, was father to Saturn and grandfather to Jupiter, who, in his turn, begat Mars, Venus, Mercury, and Apollo (or the Sun). By the mid-nineteenth century, the name "Uranus" became generally accepted. Incidentally, the 92nd chemical element, uranium, was discovered only a few years after Herschel's work, and so the element was named for the planet.

While the discovery of Uranus resulted from a pure accident, that of Neptune was a triumph for Newtonian theory and the scientific method. Careful observations in the early nineteenth century revealed irregularities in the orbit of Uranus. Two great theoretical astronomers attacked the problem, one in France and another in England. They postulated the existence of an eighth planet whose existence would perturb the motion of Uranus in just such a fashion as was observed. The paper that John Couch Adams left at the Royal Observatory on October 21, 1845, began, "According to my calculations, the observed irregularities in the motion of Uranus may be accounted for by supposing the existence of an exterior planet, the mass and orbit of which are as follows . . . " Meanwhile, and completely independently, Urbain Leverrier, in Paris, prepared a paper entitled *"Sur la Planete qui produit les anomalies observées dans le mouvement d'Uranus—determination de sa masse, de son orbite, et de sa position actuelle.*'' Each scientist explained precisely where and when to search the skies, and just what to look for. A short time later, on September 25, 1846, the Berlin observatory announced the discovery of the predicted planet. It was indeed, in the words of the director of the observatory ''the most outstanding proof of the validity of universal gravitation,'' Eventually the eighth planet was named ''Neptune,'' but not without a fight.

Still, there remained small, but certain, irregularities in the motions of the outer planets. A ninth planet was predicted by Percival Lowell in 1915, but because a powerful enough telescope had not yet been built to detect this faint and distant object, he did not live to see it discovered. Pluto was found by astronomers working at the Lowell Observatory at Flagstaff, Arizona in 1930 and announced on Lowell's birthday. Its astronomical symbol ♇ is constructed out of its champion's initials. The first two transuranic elements, which do not occur naturally, were first synthesized at the Berkeley cyclotron. Coming in the periodic table, just after uranium, it is understandable that they were named neptunium and plutonium, after the last discovered planets. Today, the art of growing new elements has developed ever further, and there are now *fourteen* known transplutonic elements. The game of naming chemical elements has lost its charm, and the last few elements simply have numbers.

Much more could be said about the Solar System. The photos returned by the Voyager spacecraft of Jupiter and Saturn and their moons would bring tears of ecstasy to Kepler's eyes, and do just that to lesser mortals. Uranus and Neptune are found to have rings much like Saturn's. It has been speculated that the Sun has a dwarf companion star called Nemesis, whose passage through the Solar System each 26 million years causes death and destruction on Earth. Nemesis, it is said, was responsible for the extinction of the dinosaurs, and it is due to return in another 10 million years. Eternally fascinating as our own solar system remains, it is very much a provincial issue. Let us turn our attention to greater questions.

The Milky Way, Galileo saw, consists of a vast number of stars spread out across the sky in one continuous and encircling band. It was reasonable to suppose that the Sun is but one star of many comprising a vast disk-shaped aggregation. Indeed, our Milky Way (called the Galaxy, with a capital G) consists of about 10^{11} stars in a disk about 80,000 light years across and about 6,000 light years thick. Just as the Earth is not at the center of the Solar System, our sun is nowhere near the center of the Galaxy: Sol lies about 30,000 light years away from the center, and somewhat "north" of the midplane of the galactic disk.

Almost every star visible to the naked eye is part of the Galaxy. One conspicuous exception is known as Andromeda. Through a small telescope, it appears as a tiny elliptical patch of light. Using larger telescopes, it was discovered that there are thousands of such faint and fuzzy "nebulae" scattered about the sky. What could they be? Listen to the philosopher-scientist Immanuel Kant writing in 1755:

> . . . a nebula [galaxy] is not a unique and solitary sun, but a system of numerous suns, which appear crowded, because of their distance, into a space so limited that their light, which would be imperceptible were each of them isolated, suffices, owing to their enormous numbers, to give a pale and uniform luster. Their analogy with our own system of stars; their form, which is precisely what it should be according to our theory; the faintness of their light, which denotes an infinite distance; all are in admirable accord and lead us to consider these ellipitical spots as systems of the same order as our own—in a word, to be Milky Ways similar to the one whose constitution we have explained . . . We see that scattered through space out to infinite distances, there exist similar systems of stars (galaxies), and that creation, in the whole extent of its infinite grandeur, is everywhere organized into systems whose members are in relation with one another. A vast field lies open to discoveries, and observation alone will give the key.

Thus did Kant introduce the notion of "island universes," distant galaxies similar in structure and composition to our own galaxy. The nearest galaxy to ours, and the brightest, is Andromeda, and it is to Andromeda that Edwin Hubble pointed his giant 100-inch telescope on Mount Wilson. Despite the enormous distance to our neighbor galaxy, Hubble was able to make out individual stars: Andromeda is indeed a "system of numerous suns"! More than this, Hubble identified within Andromeda certain very special stars called Cepheid variables. The distance to these stars may be reliably estimated in terms of their apparent brightness. Hubble was able to show that the distance to Andromeda is more than a million light years. Andromeda is a galaxy very much like our own. Thus, this picture of our sister galaxy in Figure 3 might look much like a picture of the Milky Way taken by a bug-eyed monster from Andromeda.

Having established the "island universe" hypothesis, Hubble was yet to make a far greater discovery in 1929—what we now know as

FIGURE 3. *Andromeda, the nearest galaxy, is similar to the Milky Way.*

"Hubble's law," our first and most important clue that we live in an expanding universe. Let us attend his famous Yale Silliman Lectures delivered in November 1935:

> [Galactic] spectra are peculiar in that the [spectral] lines are not in the usual positions found in nearby light sources. They are displaced toward the red of their normal position, as indicated by suitable comparison spectra. The displacements, called red-shifts, increase, on the average, with the apparent faintness of the nebula that is observed. Since apparent faintness measures distance, it follows that red-shifts increase with distance. Detailed investigation shows that the relation is linear.
>
> Small microscopic shifts, either to the red or to the violet, have long been known in the spectra of astronomical bodies other than [galaxies]. These displacements are confidently interpreted as the results of motion in the line of sight—radial velocities of recession [red-shifts] or of approach [blue-shifts]. The same interpretation is frequently applied to the red-shifts in [galactic] spectra, and has led to the term velocity-distance relation . . . On this assumption, the [galaxies] are supposed to be rushing away from our region of space, with velocities that increase directly with distance.

According to Hubble's law, distant galaxies are speeding away from us, and there is a simple, linear correlation between the speed of recession and the distance of the galaxy. In Figure 4, we plot these quantities for five galaxies. The relation is a simple one: nearby galaxies recede slowly, distant ones speed away more rapidly. Double the distance, double the speed. The recessional speed of a galaxy is given approximately by 15 kilometers per second times its distance in millions of light years.

The constant of proportionality relating the recessional speed of a distant galaxy to its distance is known as the Hubble constant, and it is one of the most fundamental parameters of observational astronomy. It is, in a sense, the rate at which the universe is exploding, and it is very closely linked to the age of the universe. Unfortunately, it is not a precisely measured quantity, and is known with only 50 percent precision. Indeed, its current value is about a factor of 10 smaller than the value Hubble originally announced. The precise determination of the Hubble constant is exceedingly difficult. What is needed is a direct measurement of the

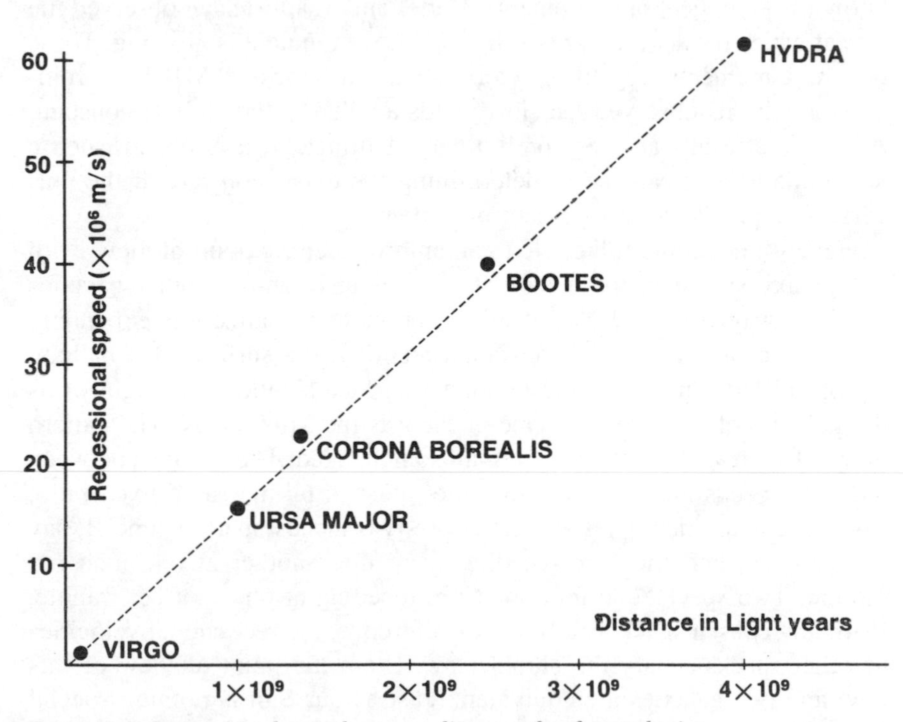

FIGURE 4. *Recessional speed versus distance for five galaxies.*

distances to distant galaxies. Most analyses are based upon the apparent brightness of galaxies. If all galaxies were equally bright intrinsically, then their brightness as seen on Earth would be a measure of their distances. Unfortunately, this is not the case, and the determination of the Hubble constant depends upon a concatenation of experimental techniques of various degrees of certitude.

Fortunately, very recent developments in radio astronomy have, for the first time, made possible a direct determination of the Hubble constant. The method, pioneered by Norbert Bartel and Irwin Shapiro at the Harvard-Smithsonian Center for Astrophysics makes use of VLBI, or very-long-base-line interferometry, in which several radio-telescopes thousands of miles apart are linked together. The analysis depends upon the presence of a recent supernova in a distant galaxy. (On average, a supernova explodes in an average galaxy about once every decade.) Bartel and Shapiro focused on the galaxy known as M101, which experienced a supernova in 1979 and lies about 60 million light years away. The supernova, now some five years old, is surrounded by an enormous expanding shell of hot gases. The velocity of expansion of the shell is known from optical measurements. Bartel and Shapiro have observed the rate at which the angular size of the supernova remnant is growing. These two measurements, together, determine the distance of M101. A comparison with its observed red shift yields a value of the Hubble constant. As yet, the results are very preliminary. Ultimately, it is hoped, such a technique will be capable of determining the expansion rate of the universe to a precision of 1 percent or better.

Have we not come full circle to an anthropocentric point of view, with our galaxy at rest in the center of the universe, and all other galaxies rushing away from us? Not at all. In order to visualize our expanding universe, consider a two-dimensional analogy, the surface of a balloon being inflated. Specks on the balloon's surface should be thought of as the galaxies of our universe. One of them is the Milky Way. The pattern of specks grows, but otherwise remains unchanged. The distance between any two specks increases, in any time interval, by the same fixed ratio. Two specks an inch apart at time A are two inches apart at time B, say one minute later: they are receding from one another at one inch per minute. Two specks one foot apart are receding at one foot per minute. With any chosen speck as a frame of reference, the recessional velocities of other specks satisfy the Hubble law. From the point of view of this law, any two galaxies are equivalent. Just as our Sun is nothing special in the Galaxy, our galaxy is nothing special in the universe. Notice that our analogy has another curious property: the surface of a balloon is finite, but it has no edges, and any one point is the same as any other. While we do not yet know whether our universe is finite or of infinite extent, we know it has no boundaries, and we devoutly believe that it looks more or less the same from any position. This last principle, which is evidently quite impossible to verify experimentally, has been elevated to dogma. It is known as the cosmological principle.

I mentioned that one thing we do not know is whether the universe is infinite in spatial extent. This is certainly an important question, and it is one scientists will answer in the not-too-distant future. The answer depends upon just what the average density of matter is in the universe. We know this number, but not yet with sufficient precision to decide upon the finiteness of the universe. The betting in the astronomical community is that the universe is indeed infinite. Assume that they are right.

If the universe is infinite, and if the distribution of stars and galaxies is everywhere the same (the cosmological principle!), we would seem to have a bit of a problem on our hands, a problem known as Olbers' paradox. According to Olbers, in an infinite universe the night sky cannot be dark. Look off in any direction and, sooner or later, your line of sight will encounter a star. Even though the stars are very far apart, there are an infinite number of them and each has a finite size. Olbers argued (correctly) that the intensity of light on earth, day and night, would be in the equivalent of 10,000 suns!

Even a child knows that this is not so. Night is dark and scary. What the child does not know is that the expanding universe resolves the paradox. For, if the universe is expanding, it could not have been expanding forever. From the observed expansion rate, we can deduce that the age of the universe is not more than, and probably near to, 20 billion years. Infinite though the universe may be, light cannot reach us from distances greater than 20 billion light years. Moreover, light from distant sources is shifted to the far infrared and is entirely ineffectual. The universe may or may not be finite, and the sky will remain dark. But it does have a beginning. Just as the universe is expanding in the future, it is contracting as one looks in the past. One day, long ago, all of Earth, all the Solar System, all of the Milky Way, and all the known galaxies would have fit within a pumpkin seed. Not a myth, this is the scientists' view of creation: First there was the Big Bang.

Life on Log Time

Consider the growth and development of a human being from the moment of conception onwards, the miracle of life. At least nine distinct stages of development may be distinguished. The fertilized egg, after about one day, begins to divide furiously, becoming a free blastocyst—a spherical body of cells travelling through the mother's reproductive tract. At six days, its wanderings are over, and it must connect itself to the wall of the uterus, becoming an attached blastocyst. After about a month, the intricate construction of the placenta is completed, and the hungry embryo begins to feed upon its mother's bloodstream. Beyond the first trimester, the person to be is usually spoken of as a fetus. The first postnatal stage of human life is infancy, followed by childhood, adolescence, and finally maturity. These nine stages of life may be displayed in the form provided in Table 1.

TABLE 1. The Nine Stages of Life.

Stage of Life	*Beginning*
1. Fertilized egg	Conception
2. Free blastocyst	1 day
3. Attached blastocyst	6 days
4. Embryo	4 weeks
5. Fetus	12 weeks
6. Infant	38 weeks
7. Child	2.5 years
8. Teenager	10 years
9. Adult	21 years

Surely it would be more enlightening to present this information in the form of a picture. Let us try with Figure 1. This illustration of the nine stages of life leaves much to be desired: most of the significant information is compressed into the very top of the picture.

The trouble with the picture is that each unit of length corresponds to a fixed period of time. It is a ''linear,'' or ''arithmetic,'' display. This would be fine and dandy for a discussion of American politics over the past two centuries, since a new president is elected every four years, like clockwork. It is not appropriate for a description of human development, where things happen much more quickly and dramatically at the beginning. (Children learn to ride bicycles, and teenagers suffer puberty, but rarely do adolescents sprout arms, legs, and livers.) Even from a psy-

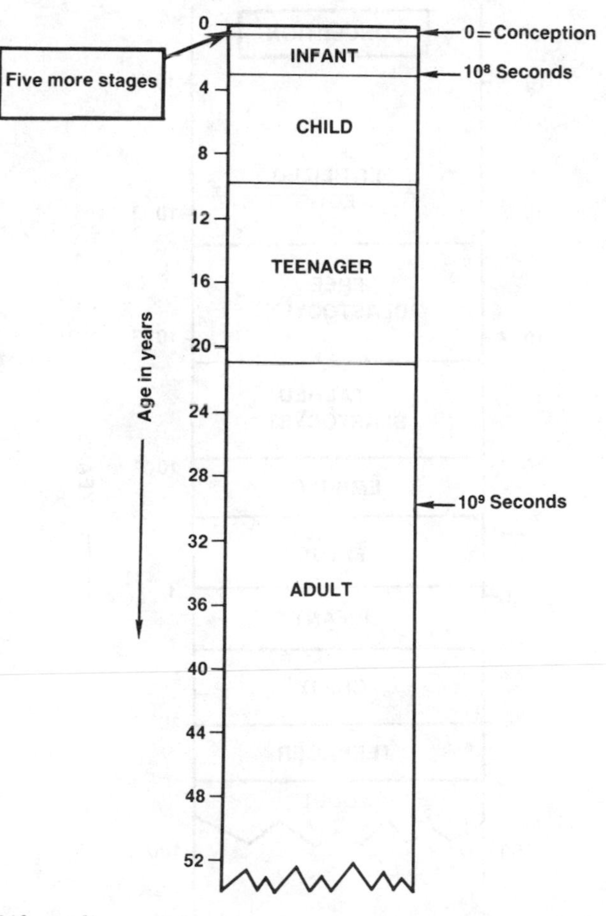

FIGURE 1. *Life on linear time.*

chological viewpoint, the linear progression of time is flawed. Time seems to pass more rapidly as we grow older. One year, to a five-year old, may be like two to a ten-year old or ten to a fifty-year old. Perhaps subjective time is measured in terms of time already spent on earth.

Clever mathematicians have invented another kind of scale, more appropriate to the situation at hand—the "logarithmic" scale, in which each unit of length corresponds to a fixed multiplicative factor. Behold in Figure 2 the panorama of human development in log time.

In the logarithmic display, the nine stages of life are well separated, and it is clear that most of the significant developments take place before birth.

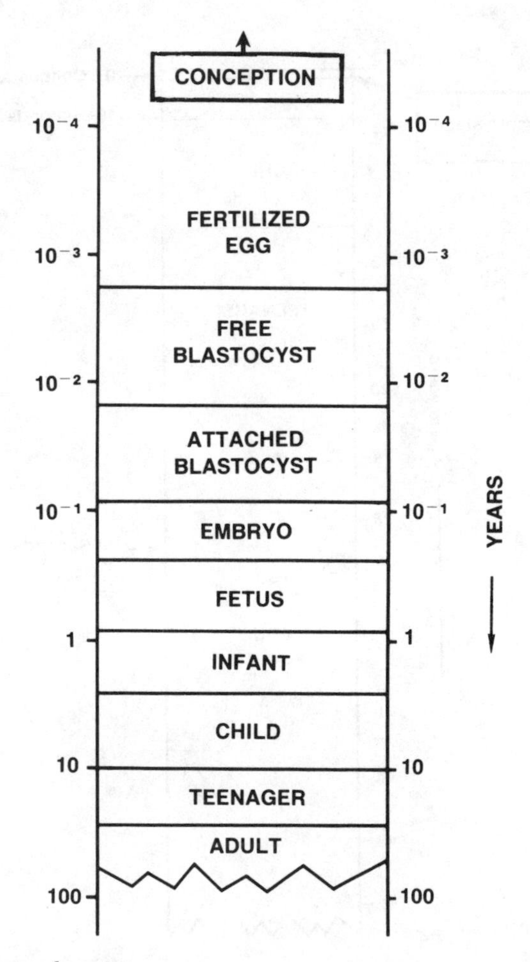

FIGURE 2. *Life on log time.*

Notice that the moment of conception has been pushed off to infinity, which may suggest the existence of even earlier stages of human development. Indeed there are: the attachment of one sperm cell to the egg, its rejection of subsequent suitors, penetration of the sperm cell into the egg, the amalgamation of parental DNA, and so on.

Childhood begins at about the hundred-millionth second after conception, or at about two years, eight months. In these terms, one's next significant anniversary occurs at the billionth second, when one is just over thirty. Only Methuselah could celebrate his next, 10^{10} second, birthday, at the ripe old age of 317 plus.

The reader may wonder what our digression on human embryology has to do with the expansion of the universe. The point is only to demonstrate, in a down-to-earth way, the power of the logarithmic scale. It is a device we shall often employ. For example, let us apply it to the birth and evolution of the entire universe—surely as anthropomorphic an analogy as may be imagined. The beginning of the universe, in the titanic explosion known as the hot Big Bang plays the role of the moment of conception (or perhaps, the little bang) and today's universe is the analog of an adult human being.

We may distinguish seven stages in evolution of the universe since it was created 20 billion years ago. The first stage is the universe in its first microsecond of existence. So hot is it that only the most fundamental building blocks of matter can survive. It is the Age of Quarks and Gluons. It was during the early part of this age that the commitment was made to matter rather than antimatter as the dominant ingredient of today's universe. As time passes, the universe expands and cools. A point is reached when quarks can combine with one another to form neutrons and protons and their antiparticles. This Age of Nucleons and Antinucleons persists from the age of a microsecond to about a hundredth of a second. At this point, while still incredibly hot, the universe has cooled enough to permit the mutual annihilation of nucleons and antinucleons. Only a few excess nucleons survive; much later, they will become the basis of matter as we know it. At this point in its development, the composition of the universe is dominated by electrons, positrons and neutrinos—the Age of Leptons, and it persists until the universe is about 100 seconds old.

At this point, a number of important things begin to happen at the same time. The net result is to permit the formation of simple atomic nuclei, like those of helium and lithium. At the end of this period, the Age of Nucleosynthesis, the matter of the universe consists of about 74 protons (hydrogen nuclei), 25 percent helium nuclei, and 1 percent of heavier atomic nuclei. The process is completed when the universe is about 1,000 seconds old. With the ending of the fourth phase of the evolution of the universe, most of the complex reorganization of matter and energy has been completed. The stage has been set for the production of the stars, galaxies, planets, and more exotic bodies that make up the heavenly panorama. However, the universe is still too hot for these objects to form. Indeed, it is still too hot for atoms to form: it is the Age of Ions, in which the universe is a hot, homogeneous, and opaque plasma. In log time, this is by far the longest of the seven stages of universal evolution. It lasts for a full ten powers of ten: from the age of a thousand seconds to 10^{13} seconds (or 500,000 years).

The Age of Ions is a time of expansion and of cooling. Nothing much of interest happens until the temperature falls to an icy 3,000°C, about half the temperature of the surface of the sun.

The point has been reached at which the electrons are captured by the ions, thus initiating the Age of Atoms. Suddenly, mysteriously, throughout the universe, and all at once, the universe becomes transparent. Light can pass freely from one point to another. The universe has become a tenuous gas of hydrogen and helium, still expanding and still cooling. Eventually, a kind of condensation occurs. Small irregularities in the giant gas cloud grow larger. Eventually, when the universe is a few hundred million years old, the primordial fluctuations have become enormous: they have evolved into the luminous heavenly bodies that still inspire wonder in that curious species living on a planet called earth. The seventh stage has begun: the Age of Stars and Galaxies. We have come, finally, to the universe of today at an age of about 20 billion years.

The temperature of our universe is now about −270°C, which is to say, only three degrees above ultimate coldness, or what scientists call "absolute zero." You may have noticed that it really is not that cold on earth, not even in a New England winter. Ah, but we are fortunate to live a mere 10^{13} cm from a modest star, the sun. Should Earth be severed from the Solar System and put halfway to nowhere, we would soon discover just how cold our universe really is. The seven stages in the growth and development of the universe are shown in Figure 3. Notice that the long Age of Ions separates universal history into two major regimes: the very early universe, involving the exotic vocabulary of elementary-particle physics and the current universe of observational astronomy. (Not to worry, by the way, if you don't know the difference between a quark and a gluon . . . we will return to these beasts presently.)

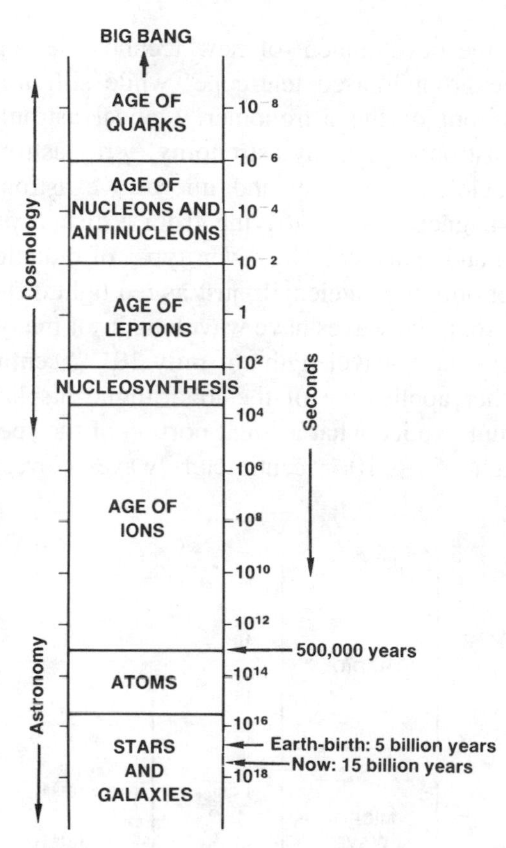

FIGURE 3. *The universe on log time.*

The early universe is the meeting point of elementary-particle physics, general relativity, and astronomy. Today, practitioners of this field (neonatal cosmologists?) are having a ball. It is a time of discovery and a time of excitement. It may in fact be possible for the scientist to understand the early and quite unobservable history of the universe. Lecturers frequently discuss the first 10^{-42} second (a tristadecillionth, if you please) since the Big Bang without being laughed off the podium. Careful now: speak of the time *before* the Big Bang, and you *will* be laughed at. Such a metaphysical concept is tabu to a physicist. It is not for nothing that we contrived our logarithmic display to push the actual moment of the Big Bang off to infinity.

The latter epoch of the universe—the story of stars and galaxies, quasars, pulsars and black holes, and the possibility of extraterrestrial life—is itself enjoying a renaissance, a time of great discoveries. This is a consequence of the development of new technologies of astronomical observation. The ground-based telescope, while still important, is no longer the only tool of the astronomer. Optical astronomy has been joined by radio astronomy, X-ray astronomy, γ-ray astronomy, infrared astronomy, ultraviolet astronomy and microwave astronomy. In each case, an electromagnetic signal arriving from a distant part of the universe is detected and analyzed. The seven types of radiation listed differ from one another only in wavelength, just as red light differs from violet light, only more so: radio waves have wavelengths of many meters, while some gamma rays have wavelengths of only 10^{-16} centimeters. Figure 4 provides another application of the logarithmic display: the electromagnetic spectrum. Notice what a small portion of the spectrum is taken up by visible light: 5×10^{-5} centimeters waves correspond to violet

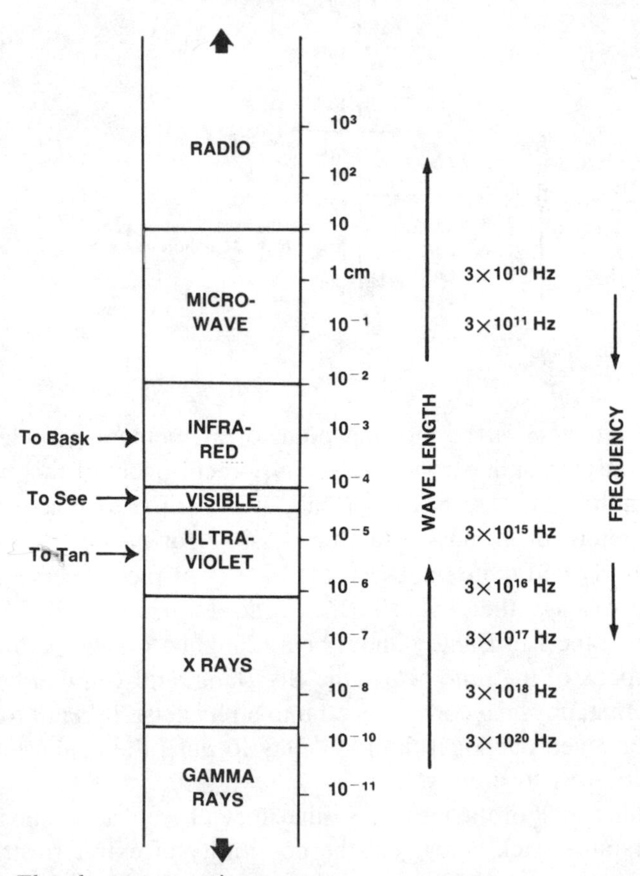

FIGURE 4. *The electromagnetic spectrum.*

light, and 8×10^{-5} centimeters is red light. Between these lie blue, green, yellow, and orange. It is not surprising that the new astronomy is so much more powerful than the old. Later, we shall discuss a few of the great new astronomical discoveries that have been made possible by our technological, curious, and generous society.

Let us stop for a moment to examine the very large numbers and very small numbers that necessarily confront us in the cosmos and within the atom. We begin with the pure, or dimensionless, numbers. An example of such a number is π, the ratio of the circumference of a circle to its diameter. Other examples are the number of planets in the Solar System (nine), the number of deadly sins (seven), the number of different poker hands (2,598,960), or the population of the United States ($\sim 2 \times 10^8$). Note the use of the symbol "$\sim$". It means "about" or "approximately." It is used when we do not know, do not care, or cannot ascertain a number with precision. The American population, in its last few digits, changes rapidly, since people are being born or dying quite often. Should the figures include resident aliens? Foreign students? Tourists? Americans living and working abroad? Dual nationals? Illegal immigrants? For the sake of general discussion, a crude estimate of the population is quite sufficient.

In science, as in such social questions, we often do not know nor do we care about the precise value of a number. A "ballpark estimate" is all that is needed. Thus, we say that the sun contains $\sim 10^{57}$ protons. On the other hand, sometimes we do care: the uranium atom, for example, contains exactly 92 electrons.

What makes a pure number pure is that its specification does *not* depend upon a specific choice of dimensional units. In the phrases "55 miles per hour," "three-minute egg," or "100-watt light bulb," the numbers bear dimensions. With conventions other than mile, minute, or watt, the numbers would change. My weight, for example, is not a pure number. In pounds it is 198, in kilograms it is 90, in stone 14.1, and in troy ounces, 2893.5668. My weight is my weight, but its numerical specification depends upon a choice of measurement standards—my weight is a dimensional number.

In Figure 5, we show a logarithmic display of ten pure numbers spanning practically four-score powers of ten. While there are a lot of people in the world, their number pales compared to the number of stars in the universe, which is itself negligible when compared to the number of possible positions in the game of chess. The largest number we introduce is the number of protons in the universe, which is equal to the number

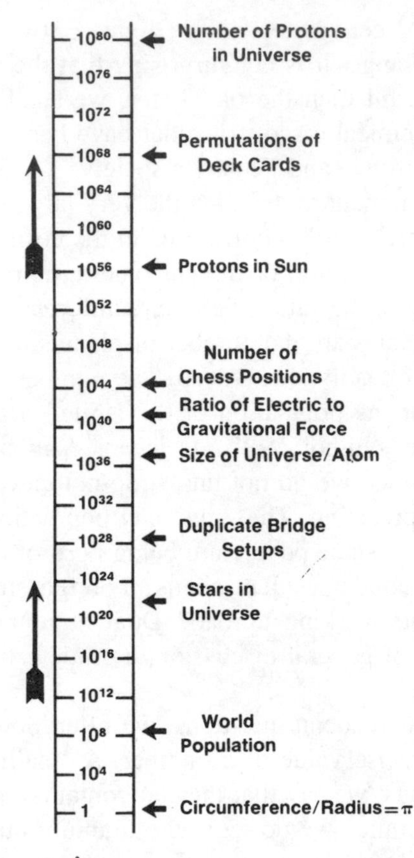

FIGURE 5. *Big pure numbers.*

of protons per average star (10^{57}) times the number of stars in the known universe (10^{22}), giving a sum total of 10^{79}.

Numbers with dimensions, like distances, times, masses, energies, pressures, velocities, and the like require the specification of a unit. Consider, for example, lengths, which Americans reckon in inches, feet, yards, and miles. Other nations, over the centuries, chose other standards of length. Table 2 provides a list of 100 units of length from a variety of cultures.

TABLE 2. Units of Length

Angstrom Unit	Braca	Centimeter	Chek
Archin	Braccio d'Ara	Chain (G.B.)	Ch'ih
Astronomical Unit	Brasse	Chain (U.S.)	Cho
Aune	Braza	Chang	Coss

Covado
Cubit
Daktylos
Diraa
Dito
Duim
El
Ell
Ella
Elle
Estadio
Fathom
Fermi
Fod
Fot
Foot (Fr.)
Foot (U.S.)
Furlong
Fusz
Guz
Hand
Hat'h
Inch
Kassabah
Ken
Kerat
Kette
Khat
Kilometer
Klafter
Kung
Kup
Latro
League
Li
Light Year
Lignc
Link
Liniya
Meter
Micron
Miglio
Mijl
Mil
Mile (Austria)
Mile (Nautical)
Mile (Norway)
Mile (U.S.)
Millimeter
Mkono
Nin
Palame
Palm
Palmo
Parsec
Pe
Pecheus
Perch
Persakh
Pharoagh
Picki
Pi
Pied
Pied de Roi
Pouce
Pu
Pulgada
Ri
Rode
Rod
Sagene
Sen
Shaku
Streep
Strich
Toisc
T'sun
Tu
Vara
Verchok
Verst
Yard
Zar
Zoll

Obviously, things appear to be getting out of hand, especially for the traveler or the merchant. It would save a lot of trouble if the different nations of the world could agree to abandon their tribal or traditional units, which are usually based upon the physiognomy of a long forgotten tyrant. As a matter of fact, most countries have done just this. All major nations of the world, with the exception of the United States, have adopted the metric system. Throughout the un-American world, the meter is the basic measure of length, and the gram is the basic measure of mass. The only remaining nations with no plan to go metric are Burma, both Yemens*, and the U.S. American engineers, trained in terms of slugs, B.T.U., Fahrenheit degrees and foot-pounds can barely communicate with their metrical colleagues in Europe and in Asia. American manufactured goods, unless they are expensively customized to the export trade, require nonmetric tools for their maintenance for repair. Can this be an effective selling point in a metric world?

American industry was once foremost in the world. Today, most of our ships, stereos, shoes, sewing machines, subway cars, TVs, calcula-

* *Author's note:* The two Yemens have since become unified.

tors, video recorders, and musical synthesizers are made abroad. Is our stubborn failure to abandon the pint and the pound, the inch and mile, merely a symptom of decline, or is it among the causes?

Scientists throughout the world use metric units. The meter, at 3.281 feet, is a bit more than a yard. The centimeter, the hundredth part of a meter, is about the width of your pinkie fingernail, or .3937 inches. One pleasing feature of the metric system is its use of certain prefixes which modify the fundamental unit by powers of ten. The *centi*meter, for example, is 10^{-2} meters, and the *deci*liter is 10^{-1} liters. Table 3 presents a list of other useful prefixes, which are equally spaced in logarithmic measure by three powers of ten.

TABLE 3. Metric Prefixes

Name	*Multiplies By:*	*Comes From:*	*Meaning:*
tera	10^{12}	Latin	Monster
giga	10^{9}	Latin	Giant
mega	10^{6}	Greek	Great
kilo	10^{3}	Greek	Thousand
milli	10^{-3}	Latin	Thousand
micro	10^{-6}	Greek	Small
nano	10^{-9}	Greek	Dwarf
pico	10^{-12}	Italian	Small
femto	10^{-15}	Danish	Fifteen
atto	10^{-18}	Danish	Eighteen

While not all of these prefixes are of everyday currency, one begins to hear of national power consumption in *terawatts,* or of the Japanese challenge in *picosecond* technology, or of the cost of yet another ineffective weapon system in *gigabucks.* (A gigabuck per megadeath, in the nuclear holocaust, comes to merely $1,000 per funeral.)

In Figure 6 are shown various representative distances in logarithmic display. The world of human affairs lies in the vicinity of 10^2 cm, and the plot extends over more than 45 powers of 10, from the smallest distances we have investigated at our giant accelerators, to the most distant known realm of the heavens.

Having considered time and distance, let's turn our attention to speed or velocity. Everyday units of speed are kilometers per hour, or miles per hour, or knots. Our choice of unit will be centimeters per second. The U.S. highway speed limit, 55 miles per hour, corresponds to about 2,400 cm/s, while the almost record human speed of 10 seconds for the

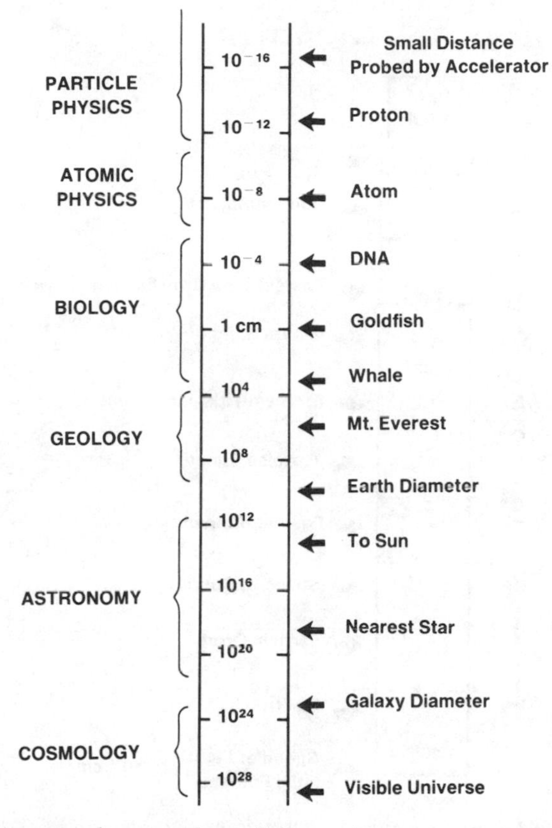

FIGURE 6. *Distances large and small in centimeters.*

100-meter dash corresponds to 1,000 cm/s. Various representative speeds are shown in Figure 7 in logarithmic display. The slowest speed we could readily visualize is the rate of growth of the human fingernail: about an inch per year, or 10^{-7} cm/s. Coincidentally, this is about the same as the velocity of continental drift. Slow though this speed is, the continents will move a thousand kilometers, quite enough to screw up the map of the world, in 30 million years, a mere 1 percent of the age of the earth.

Now to really fast speeds. Military jets travel a few times faster than sound at a speed of perhaps 10^5 cm/s. Escape velocity from Earth is about 10 times greater. The speed at which the Earth revolves about the Sun is 3×10^6 cm/s, and our entire solar system is rushing through the universe at the speed of 4×10^7 cm/s, relative to the remnants of the Big Bang. Electrons in atoms move at about 3×10^8 cm/s, and those that produce the picture in your television set move 10 times faster still,

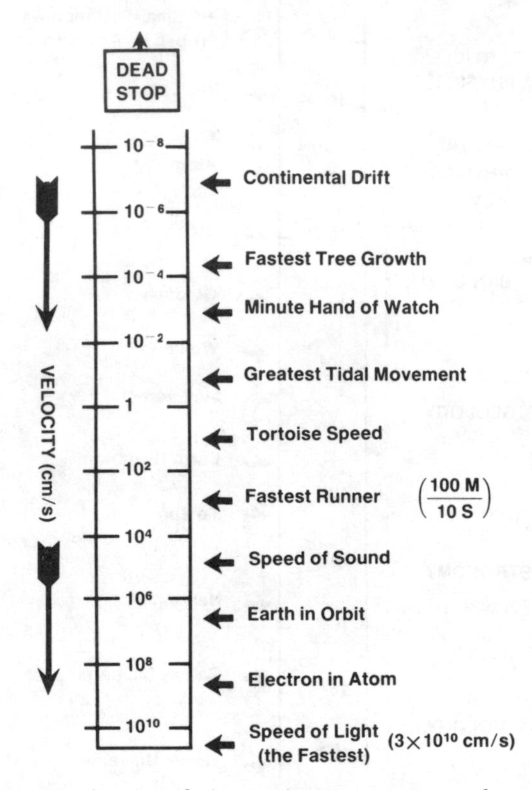

FIGURE 7. *Representative speeds in centimeters per second.*

at 3×10^9 cm/s. Finally, we come to light itself, which holds all possible records for speed. So fundamental is the velocity of light that it has its own universal symbol, c. Its value is 2.998 (to all intents and purposes, 3) times 10^{10} cm/s, or in patriotic units, 186,000 miles per second. According to the special theory of relativity, a generally accepted part of contemporary science, the velocity of light is a limiting speed. No object, message, or signal of any kind can travel faster than light. There is not and will never be a faster-than-light drive.

The Number Game

An infant, exploring its fingers and toes, learns to count and discovers the integers. It is a momentous discovery, for there are an infinite number of integers, far more than anyone can count. But, the result of this count, the number of integers, is a meaningful mathematical concept. This infinite number is called "aleph-null" by mathematicians. Infinite numbers, as we shall see, are quite peculiar beasts.

Consider the number of integers which are divisible by 17. Most people would guess that there are fewer of these than there are integers. After all, not all integers can be divided by 17. However, the number of such integers is also equal to aleph-null. Two sets are said to be equal in number if a faithful correspondence can be established between the members of one set and the members of the other. In a large room filled with dancing couples, it is easy to tell that the number of men equals the number of women. It is not at all necessary to count them. Here is an analogous correspondence between the set of all integers and the set of integers divisible by 17:

$$1 \leftrightarrow 17$$
$$2 \leftrightarrow 34$$
$$3 \leftrightarrow 51$$
etc.

Every integer is coupled to a unique multiple of 17, and conversely. The two sets are equal in number.

An integer which is divisible by no other integer but one is called a prime. Seventeen is a prime, as are two, three, and five. Let us whet our logical skills by proving that there are an infinite number of primes. Our method of proof is called *reductio ad absurdum.* We shall assume the contrary of our assertion, and deduce a contradiction. This means that the contrary of our assertion must be false, so that the assertion itself must be true. Suppose that there are only a finite number of primes. Multiply them all together and add one. The resulting number is evidently not exactly divided by any of the primes on our list. It follows that it is itself a prime. This is a contradiction, since this number was not in our original set which was supposed to contain all the primes. Thus, there are an infinite number of primes. More precisely, the number of primes is aleph-null.

Children, by dividing a sweet among their friends, discover fractions. These are obtained by dividing one integer by another. The set of all fractions is called "the rational numbers." Surely, it would seem that the number of fractions is considerably larger than the number of integers. There are, after all, an infinite number of fractions lying between zero and one. Indeed, there are an infinite number of fractions lying between any two distinct fractions. Nonetheless, the number of rational numbers is merely aleph-null. There are just as many fractions as there are integers. We prove this by exhibiting a faithful correspondence between the integers and the fractions:

$$
\begin{array}{lll}
1/1 & \leftrightarrow & 1 \\
1/2 \ 2/1 & \leftrightarrow & 2 \ 3 \\
1/3 \ 2/2 \ 3/1 & \leftrightarrow & 4 \ 5 \ 6 \\
1/4 \ 2/3 \ 3/2 \ 4/1 & \leftrightarrow & 7 \ 8 \ 9 \ 10 \\
1/5 \ 2/4 \ 3/3 \ 4/2 \ 5/1 & \leftrightarrow & 11 \ 12 \ 13 \ 14 \ 15 \\
\text{etc.} & & \text{etc.}
\end{array}
$$

We have listed the fractions in order of the sum of the numerator and denominator. There is only one fraction for which this sum is two, two for which it is three, three for which it is four, etc. In the above correspondence, the fraction two-thirds is linked to the integer eight. Every fraction is linked to a different integer, and conversely. The number of fractions is thus aleph-null.

There are many numbers which may not be expressed as fractions. Such numbers are called irrational. One encounters these numbers in the search for solutions to algebraic equations. Even such a simple equation as $x^2 = 2$ has an irrational solution. With our hand calculator, we learn that the square root of two is 1.4142136. Of course, this is not the exact answer. The decimal expansion continues *ad infinitum* and it never repeats itself. (On the other hand, the decimal expansion of a rational number must repeat itself. Thus $1/7 = .142857142857\ldots$). It is straightforward to prove that the square root of two is indeed an example of an irrational number.

It is obvious that there are very many irrational numbers, like the square root of three or the cube root of four. The integers, the fractions, square roots, cube roots, and the like are all included in a set called "algebraic numbers." An algebraic number is defined to be a solution to an algebraic equation, like

$$x - 5 = 0$$
$$x^2 - 2 = 0$$
$$2x^5 - 3x + 1 = 0$$
etc.

Clearly, there are an infinite number of different algebraic equations. However, they may be put into a systematic list. Since each algebraic equation has only a finite number of solutions the algebraic numbers may also be put into a systematic list or correspondence with the integers. The number of algebraic numbers is therefore equal to aleph-null. There are just as many algebraic numbers as there are integers!

Have we left something out? Are there any numbers which are not algebraic numbers? Indeed there are. They are called "transcendental numbers." An example is the famous number pi (3.14159 ...) which denotes the ratio of the circumference of a circle to its diameter. There are an infinite number of transcendental numbers, and we have finally come to an infinity that is larger than aleph-null. And this I will now prove, using once again the powerful method of *reductio ad absurdum.*

Assume that the numbers between zero and one can be put into correspondence with the integers. This would be the case if there were aleph-null of them. We shall show that our assumption leads to a contradiction. Each of the numbers on our list can be expressed by an infinitely long decimal expansion. The conjectured correspondence might look as follows

$$1 \leftrightarrow .01248\ldots$$
$$2 \leftrightarrow .99216\ldots$$
$$3 \leftrightarrow .55158\ldots$$
$$4 \leftrightarrow .01306\ldots$$
etc.

In this list, every number between zero and one is supposed to appear at some point. I shall demonstrate that this simply cannot be. I shall show how to construct a number x which cannot appear on the list. It is remarkably easy to do. The number x should have a decimal expansion with the following properties. Its first digit must differ from the first digit of the first number on the list. Its second digit must differ from the second digit of the second number on the list. And so on. Thus, x is not equal to the first number, nor to the second, nor to any number on the list. x is simply not on the list. This contradiction proves that there cannot be a list of all the numbers between zero and one. The number of numbers between zero and one is larger than aleph-null. This infinite number is called aleph-one. There is a hierarchy of infinities after all.

From integers, to fractions, to algebraic numbers, to transcendental numbers. Finally, we have elucidated all of the numbers corresponding to the points on a line. From the intuitive notion of integers we have been led to the sophisticated concept that the mathematicians call "real numbers." This is the very beginning of the enormous, exciting, and open-ended mathematical discipline called "analysis." Without analysis, there could be no calculus nor differential equations. Without these powerful mathematical tools, our technological society could not have arisen.

Welcome to UBS

The Universal Broadcasting System brings you an exciting variety of radio programs coming from just about anywhere in the universe. Daytime programming originates in our very own solar system with transmitters located at the Sun. Don't miss the next solar flare, a display of fireworks 40 times larger than the Earth itself, simulcast in both visible and radio frequencies. Evening hours are reserved for our more distant affiliates. Follow the unending saga of the birth and death of stars in our Milky Way. For the serious listener, there will be a panel discussion with more than 50 participating galaxies on the subject of the mysterious dark mass of the universe. Quasars and radio galaxies are the superstars of our monster show, while Nobel Laureates Arno Penzias and Robert Wilson sermonize on the creation of the universe in the hot Big Bang. Last, but not least, our talk show will solicit calls from intelligent extraterrestrials with interesting things to say about life elsewhere in the Galaxy. Still jaded? Turn your dial to our sister stations in the infrared to hear the latest news about the existence of other planets of other stars. But first, a word from our sponsor about what radio astronomy is all about, how it came to be, and why it is important.

The cool green hills of Earth and its abundant seas make the planet a true garden of Eden for its most recent tenants, we humans. Its atmosphere, too, plays many quiet but essential roles. Its oxygen is the life and breath of all of the complex life forms on Earth and the key to the operation of mankind's first and foremost tool, fire. Moreover, the air and its winds provides an effective transport system for water, bringing it (as rain and snow) thousands of miles from the saline oceans to the fertile lands. Without air, our planet would be dreadfully hot in the day and frigid at night. This insulating mechanism, known as the ''greenhouse'' effect, depends upon absorption of infrared waves (heat radiation) by carbon dioxide in the air. Furthermore, our atmosphere shields us from dangerous short-wavelength forms of light: carcinogenic ultraviolet radiation, dangerous X rays, and mutagenic cosmic rays. As an added luxury, the ionosphere is a reflector of certain radio waves, a fact which makes possible long-range ship-to-shore and airplane-to-land communication.

The kind hospitality of planet Earth to our mundane activities is caused, in large measure, by the insubstantial, but oh-so-important, air about us.

According to ancient Egyptian legend, Ra, the lord of the Sun, begat Shu, the air god. Shu hurled the young goddess Nut into the heavens where she remains, goddess of the sky, forever separated from her adoring husband Geb, god of the Earth. Thus did Shu succeed his father to become king of the Earth. It is, as we have seen, an honor well deserved, if not for the mythological deity Shu, then for the blessed air itself.

Because we are swathed in our protective atmosphere, only certain wavelengths of electromagnetic radiation pass freely from space to us. Indeed, there are just two great windows through which the earthbound observer can peer at the universe, which are shown in Figure 1. Radiations with wavelengths outside of the windows are absorbed by the atmosphere before they can be detected on Earth. The smaller window on the left includes all of the visible spectrum as well as a little bit of the near-ultraviolet (to tan by) and of the near-infrared (to bask in). It is through this small window that the sun illuminates the Earth and powers the great photosynthetic engine of its crops. It is also through this small window that we have discovered the splendors of the night sky—first with the unaided eye, then with increasingly sophisticated telescopes and their accompanying photographic and spectroscopic accessories. Few would have dreamed, a century ago, of the astonishing discoveries that were to be made when we turned our attention to the remarkable and complementary view in the second window, in the range of wavelengths from about one centimeter to about 30 meters.

Indeed, the very concept of electromagnetic radiation did not exist until the brilliant synthesis of the laws of electricity and magnetism was achieved by James Clerk Maxwell in 1864. According to his theory, light itself is a phenomenon of pure electricity and magnetism—light is an electromagnetic wave of very short wavelength. Larger waves could exist so far as the theory was concerned, but they had never been seen in the laboratory in Maxwell's day. Heinrich Hertz, in Germany, set about to

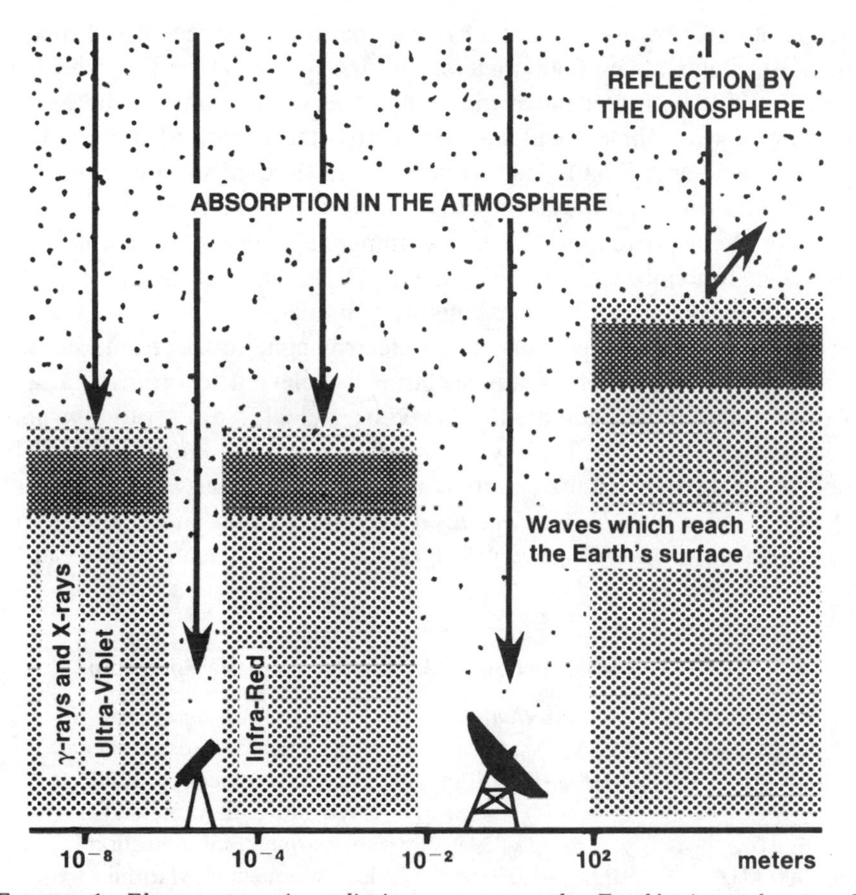

FIGURE 1. *Electromagnetic radiation penetrates the Earth's ionosphere and atmosphere through two windows or bands of wavelengths: light (the smaller window on the left) and radio waves.*

produce and detect long-wavelength electromagnetic radiation—radio waves. His apparatus consisted of a spark-gap powered by a high-voltage source placed at the center of a parabolic mirror. A second spark-gap, connected to a simple loop of wire (the antenna), was placed five feet away, within its own parabolic reflector. Zap! A spark is generated in the first spark-gap, and, lo and behold, a smaller spark is produced at the receiver. Electromagnetic waves had been produced, had sped from the transmitter to the detector, and had caused the second spark. Hertz discovered radio waves in 1888.

Guglielmo Marconi, sensing a commercially feasible development, strove to improve upon Hertz' merely pedagogical demonstrations. In a series of experiments, he sent a radio signal for a distance of 10 meters, then 300 meters, then 3,000 meters, and finally, across the English Channel. In 1901, with somewhat more sophisticated equipment, the letter *s* (dot-dot-dot in Morse code) was sent across the Atlantic from Cornwall to Newfoundland. Finally, and triumphantly, Marconi succeeded in sending a radio-telegraph message across the world, from Wales to Australia in 1918. Soon thereafter, the first commercial radio station was established in November 1920, call letters KDKA, broadcasting from Pittsburgh, Pennsylvania. The idea caught on immediately.

Consider the radio portion of the electromagnetic spectrum, summarized in Table 1. Much of this spectrum is relevant to ordinary affairs. At home and in bed, you are awakened by a clock radio set to a *medium frequency* AM station. Thaw your frozen danish with *microwaves*. Telephone the taxi dispatcher, who contacts your cab at *high frequency*. Watch a bit of *very-high frequency* TV while waiting and listen to your earphone *ultra-high frequency* FM radio during the cab ride. Board a

TABLE 1. Radio Portion of the Electromagnetic Spectrum

Frequency	*Wavelength*	*Designation and partial list of frequency assignments*
10–30 kHz	30,000–10,000 m	Very low frequency: Radio navigation, maritime and submarine communication
30–300 kHz	10,000–1,000 m	Low frequency: Maritime communication, maritime mobile satellite, intersatellite
300–3000 kHz	1,000–100 m	Medium frequency: AM radio, amateur radio
3-30 MHz	100–10 m	High frequency: Short-wave radio, amateur satellite, citizens' band radio, taxi dispatchers
30–300 MHz	10–1 m	Very high frequency (VHF): VHF television, FM radio, radio astronomy
300–3,000 MHz	1 m–10 cm	Ultrahigh frequency (UHF): UHF television, radar, synchronous satellites, cellular telephones
3,000 MHz–300 GHz	10 cm–1 mm	Microwaves: Aeronautical radio-navigation satellites, radio astronomy, microwave ovens

radar-equipped jet which navigates with the assistance of *low-frequency* instruments. In an hour, you have made use of practically the entire radio spectrum, employing electromagnetic radiation. (Of course, visible light also played a role; otherwise your cab driver would never have found the airport.)

Karl Jansky, an American radio engineer, worked at Bell Telephone Laboratories in New Jersey. In 1932, he was the first human being to listen to radio signals coming from the stars. Jansky had been studying background static which often disturbs radio listening during storms. Even in the best weather, Jansky noticed a background hiss which became intense just once a day. But, this strange signal appeared four minutes earlier from day to day. Jansky realized that this curious radio noise was correlated with the motion of the Milky Way across the sky. The signal was coming from outside the Solar System, from the direction of the center of the Galaxy!

Scientists did not seem to take Jansky's work as seriously as did science fiction writers and their fans. During the 1930s, the lone amateur astronomer, Grote Reber, pursued the infant discipline of radio astronomy. An engineer from Illinois, he built a 30-foot radio telescope in his backyard to quietly study the sky. One of Reber's remarkable observations was the discovery of a pointlike source of radio waves from space, an object in the constellation of the Swan, an object now known as Cygnus A.

Radar is an electronic system using radio waves to detect objects that are invisible to the eye. Radar devices can detect the position of an object, its distance from the observer, its speed and direction of motion, and sometimes its size and shape. The word itself is acronymic for "*R*adio *D*etection and *R*anging." While crude radio detection devices were first used in the early thirties, the invention of modern radar dates from the military necessities of the Second World War. It was the development of radar technology that made possible the explosive post-war growth of radio astronomy as well as the defeat of the Luftwaffe.

By 1954, about a hundred radio sources in the sky had been observed. Few of them could be matched up with optical counterparts that could be seen in telescopes. Many astronomers thought that those sources corresponded to otherwise invisible "radio stars" within our galaxy.

When the location of Cygnus A in the sky was pinned down by British radio astronomers, the giant 200-inch Palomar telescope could be brought to bear. Sure enough, a peculiar disturbed galaxy was found to coincide in position with the radio source. Figure 2 the Palomar image

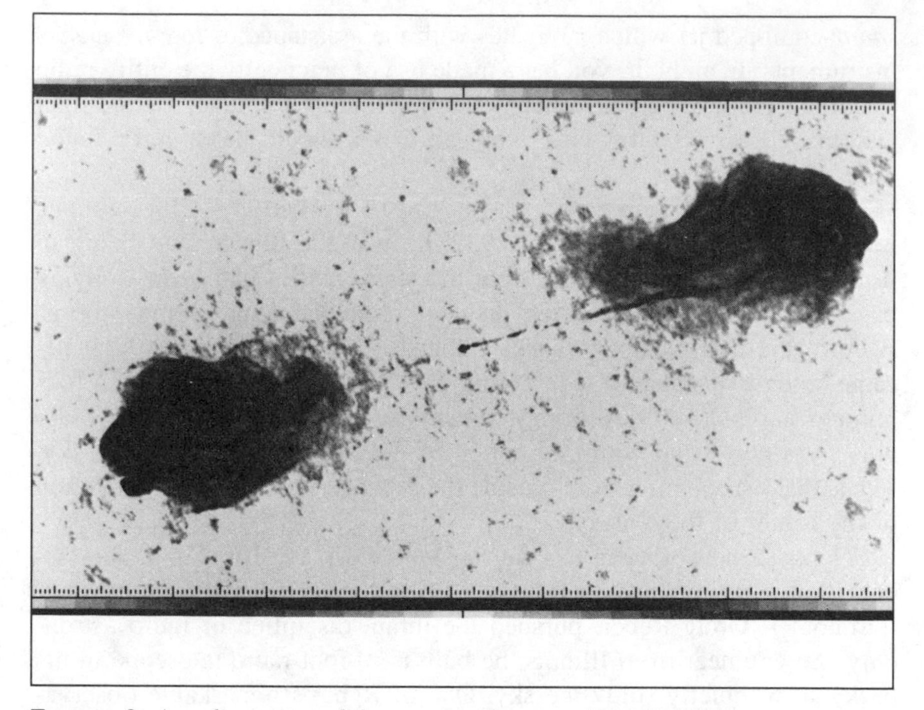

FIGURE 2. *A radio image of Cygnus A. (Courtesy NRAO/AUI.)*

shows a radio image of Cygnus A. The two radio lobes are located to either side of the central Galaxy. The radio sources are believed to consist of immense clouds of ionized plasma, each of them larger in extent than the Galaxy itself.

Once a visible object is identified with a radio source, it becomes possible to determine how far the object is from Earth. The characteristic spectral lines of the optical source are shifted to the red by an amount depending upon its recessional velocity. According to Hubble's law, the velocity is proportional to the distance. When the red shift of the central galaxy of Cygnus A was determined, the fun really began. While Cygnus A is the second brightest radio source in the sky, red-shift measurements showed that it lies at the incredible distance of a *billion* light years from Earth. The only way such a beacon can be so intense and yet so distant is for it to be an extraordinarily powerful source, sending out about 10^{38} watts of radio power. It is a radio source some 10-million times more powerful than our own galaxy. Indeed, Cygnus A generates more energy in radio waves than a typical large galaxy does in the form of visible light. It is the first-to-be-discovered member of a large class of new and bizarre beasts of the cosmos: a radio galaxy. In the decades since its discovery, many other radio galaxies have been discovered. Today, several thousands of them have been cataloged.

We do not know exactly what a radio galaxy is, but it is clear that its energy requirements are prodigious. In one popular theory, the radio galaxy is produced by an enormous spinning black hole within the central Galaxy. A black hole is an object so heavy, yet so small that nothing, not even light itself, can escape from it. We can begin to understand what a black hole is if we consider the concept of "escape velocity." A bullet shot into the air does not escape from Earth because it is moving too slowly. Escape velocity from Earth is about 11 kilometers per second, 10 times faster than the speed of the fastest jet. It is the speed that a spaceship or a satellite must achieve to escape from Earth's gravity. In the case of the Sun, escape velocity is much greater, about 600 kilometers per second. Suppose, somehow, we could compress the Sun into a ball only three kilometers in radius. (Of course, *we* can do no such thing. In other parts of the galaxy these things can and probably do happen.) Escape velocity from a body of solar mass and three-kilometer radius is the velocity of light. Since Einstein said (and we believe) that no object can move faster than the velocity of light, we have produced a trap into which it is easy to enter, but absolutely impossible to leave: it is a black hole.

Imagine, in the very center of a radio galaxy, a massive black hole which may have been produced by the collisions and coalescence of many stars in the crowded galactic interior. The black hole is surrounded by a vast and growing disc of accreted material upon which the ravenous black hole will feed and grow larger, ultimately perhaps to ingest the entire heart of the galaxy. As the black hole consumes the stars about it, a tremendous amount of energy can be released. Somehow, this energy is guided outwards in two oppositely directed beams in order to power the great radio beacons of Cygnus A and other radio galaxies.

Some astronomers believe that the creation of a black hole at the center of a galaxy is quite commonplace. There even may be a "small" black hole in the center of our Milky Way, one that has, thus far, consumed at most a few million stars. A black hole is insatiable. One day, in the distant future, our black hole will become so bloated that it will transform our own galaxy into a radio galaxy. Whether or not such an occurrence would make Earth uninhabitable is unclear, but it would cer-

omy) quite impossible. To ensure against such a possibility, you could have your home wired for cable TV.

Part of the science of botany is the patient and systematic study of the Earth's flora, compiling lists of different varieties, arranged in an orderly way. Compilation and classification are important aspects of many other sciences such as zoology, entomology, and mineralogy. The same is true of astronomy. An early, but important, list of strange bodies in the heavens is the Messier catalog, produced in 1781. It contains 103 faint and fuzzy objects (then called nebulae), neither stars nor comets. Some, like M5, are giant clusters of stars within our own galaxy. The Crab Nebula, first on Messier's list and known therefore as M1 is a remnant of the violent supernova which amazed Chinese astronomers in 1054. Another, M31, is our sister galaxy Andromeda. By 1895, the Messier catalog was supplanted by the New General Catalog with some 15,000 different entries. Since there are too many interesting objects in the sky for each to have its own name, astronomers must employ catalog numbers. The Crab Nebula, for example, being the 1,952nd entry in the New General Catalog, is sometimes referred to as NGC 1952, rather than M1.

In 1959, the new science of radio astronomy took a giant step towards scientific maturity with the publication of the Third Cambridge (3C) Catalog, consisting originally of a list of 471 radio sources. This catalog is in common use today.

Many of the radio sources in the 3C catalog correspond to known objects within our own galaxy. In the radio spectrum, the Crab Nebula is known as 3C144, for example. Other entries correspond to the exploding radio galaxies we have spoken about. The great radio source Cygnus A, first observed by Reber, is called 3C405. Several of the more peculiar galaxies in the original Messier catalog have been seen by radio telescopes and identified as radio galaxies. Shown in Figure 3 are M87 (known in radio as 3C274) and M82 (or 3C231). Clearly both galaxies are highly disturbed; one displays a vast jet of expelled material, the other shows huge filaments of hot gas erupting from the center of the galaxy. Both systems are powerful sources of radio waves and X rays.

Still, a number of radio sources could not be accounted for as radio galaxies or as unusual stars in our galaxy. Their optical counterparts could not be found. A breakthrough occurred in 1962 at an Australian radio telescope. Cyril Hazard and his collaborators set out to observe the occultation of the radio source 3C273 by the Moon. This procedure enables the radio astronomer to pin down the location of the source in the

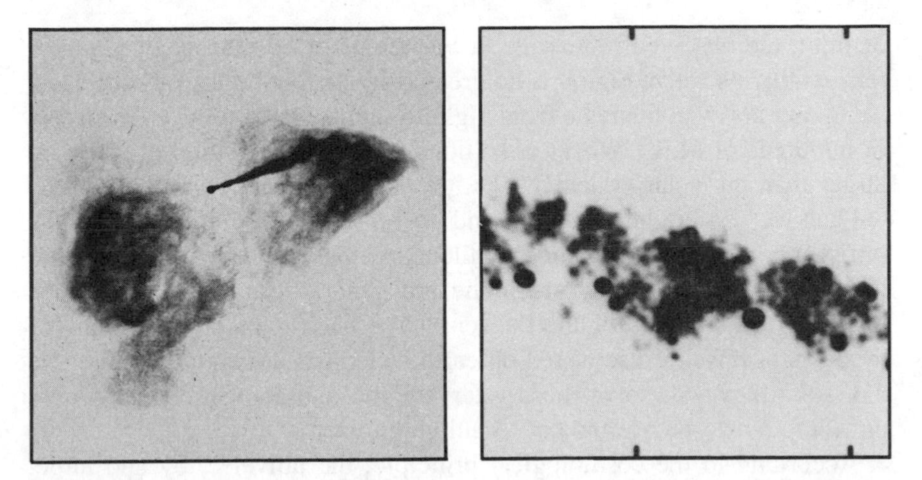

FIGURE 3. *Two powerful sources of radio waves and X rays: M87 (left) and M82. (Courtesy NRAO/AUI.)*

ables the radio astronomer to pin down the location of the source in the sky with great precision. They were able to find the star which is the visual counterpart of 3C273. From its optical spectrum, Maarten Schmidt of Caltech was able to ascertain its red shift, and could then show that it lay at a distance of about 500 million light years. And yet, the visual image of 3C273 is not all that faint. In order to reconcile its distance from us with its apparent brightness, we must conclude that it is an extraordinary source of power. Its absolute brightness must be more than 100 times that of the brightest known galaxy. Soon, 3C273 was joined by hosts of other similar objects, and they became known as "quasars" or "quasistellar objects." The most powerful known quasar, 3C279, radiates 100,000 times the power of our own galaxy.

Quasars, being the brightest objects in the sky, can be seen at the furthest distances. Some have red shifts so great that they must lie at distances of almost 20 *billion* light years. The light we see today originated at an early time in the history of the universe when it was young and surely very different from what it is today. Quasars may have been created out of the hot Big Bang even earlier than stars and galaxies, perhaps when it was only a billion years old.

Galaxies are rather large objects, generally some tens of thousands of light years in extent. Quasars, which are often far brighter than galaxies, are known to be relatively small, compact objects, since their radio and optical signals fluctuate wildly over time. Significant changes can occur in a period of weeks. Since no signal can travel faster than the velocity of light, the observed time scale of variations of luminosity of a quasar tells us that its active region is no greater than a few light weeks. Indeed, some quasars even fluctuate from night to night. They produce the energy of hundreds of Milky Ways (or trillions of stars) in a region of space no larger than our solar system!

Quasars—so bright, so small, and so far away—are among the most intriguing beasts of the cosmos. While many theories have been proposed to explain just what they are, none are generally accepted. But since many of the wonders of the heavens have been explained in terms of the laws of physics discovered on earth, scientists are confident that one day soon they will solve the mystery of the quasars. Surely they were put there not to puzzle us, but to enlighten us.

According to the cosmological principle, the universe, by and large, looks the same, no matter where the observer is located. So in constructing a map of the observable universe, we may put ourselves smack in the middle. The map in Figure 4, actually a slice through the three-dimensional universe, is shown as a circle of finite radius. The radius of the observable universe is about 20 billion light years, or 2×10^{28} cm. This is because light, or radio signals, have had only 20 billion years to travel since the universe was born. The size of our map is growing with time, at the rate of one light-year per year. As we grow older, we can see further. The map is a circle (a cross section of a sphere) as viewed by any observer anywhere in the universe and in any state of motion. There is no preferred direction; the universe is homogeneous and isotropic in the large. True, an intelligent observer upon a distant planet would see a different arrangement of stars, and would imagine different constellations. Yet, the big picture would be much the same. I have taken liberties with the scale of the Milky Way. In true proportions, it would be the size of a microscopic virus, and our solar system within it would be smaller than an atomic nucleus. The universe is immense.

In our highly schematic map, the universe is divided into five concentric shells. In the very center is the Solar System, including the seven heavenly bodies of the ancient as well as the additional planets, satellites, asteroids, and comets since discovered. Its radius is roughly a thousandth of a light year. The second shell contains the Milky Way, with a radius of about 40,000 light years. (Earth, of course, is not at the center of the Solar System, nor is the Sun in the center of the Galaxy.) Practically every star that the unassisted eye can see in the heavens lies within the second shell. The third shell contains the galaxies and extends out to a distance of a billion light years. There are probably many galaxies even

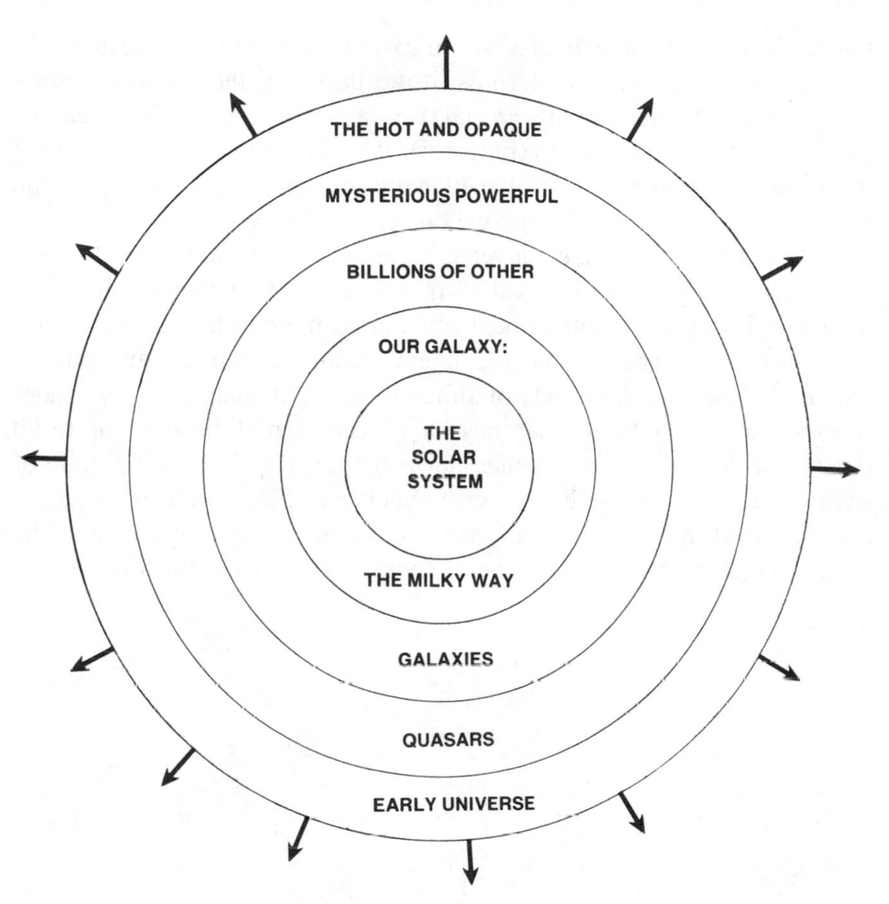

FIGURE 4. *The universe according to us.*

further away, but they cannot be seen, even with our most powerful instruments, because of their great distance from us. There are about as many galaxies in the visible universe as there are stars in the Milky Way: about a hundred billion of them. The galaxies are *not* scattered at random throughout space. They form groups or clusters of many galaxies. The Milky Way, for example, is a member of what is known in the trade as a "poor irregular cluster" and is called, affectionately, the Local Group. It is a few million light years across and includes two good sized galaxies (The Milky Way and Andromeda) together with about a dozen dwarf galaxies. The nearest rich cluster to us, about 60 million light years away, is the Virgo Cluster containing hundreds of galaxies. Not only do galaxies form clusters, but clusters of galaxies form superclusters. Virgo and our Local Group are parts of the Local Supercluster, an entity hundreds of millions of light years in extent. In addition to clusters and superclusters, there are great voids in the universe, the ultimate deserts of space in which there are no galaxies at all in regions hundreds of millions of light years across (Figure 5). Why are the galaxies distributed in this odd way, giving a peculiar filamentary structure to the large-scale structure of the universe? This, we do not yet know.

The fourth shell of the universe is populated by the powerful and mysterious compact sources called quasars. In this nether reach of the universe, billions of light years from Earth, mere galaxies are too far away to be seen. The only visible objects in this domain are the quasars, which can put out thousands of times more light than galaxies. These objects are so very bright that they have been seen at distances up to 90 percent of the way to the ultimate horizon from which the travel time of light equals the age of the universe. Such an object, were it located a few thousand light years from Earth, would be as bright as the sun. The light from distant quasars has been enormously shifted toward the red.

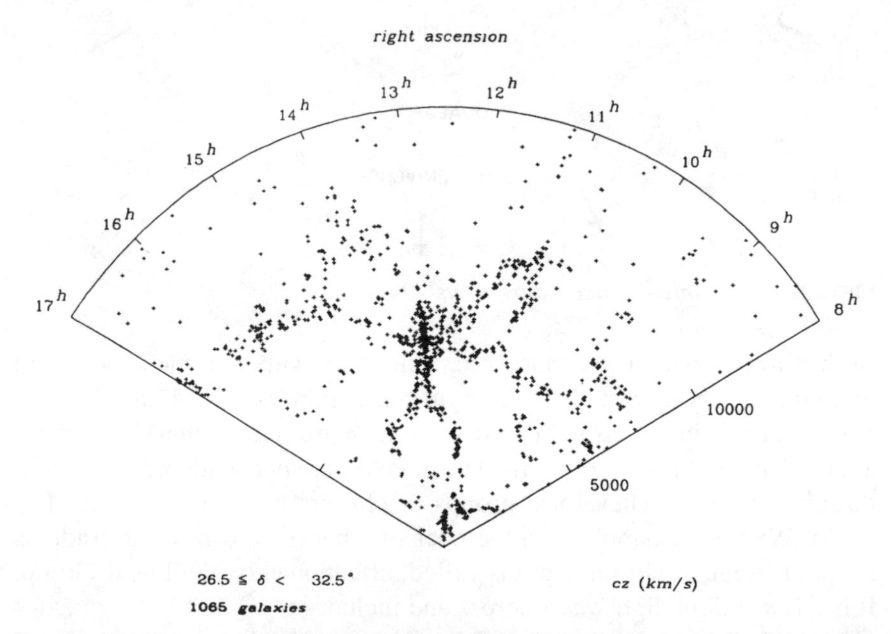

FIGURE 5. *A cross section of the universe showing the large-scale, filamentary structure. The crosses indicate the locations of galaxies found within a portion of the universe shaped like a pie slice.* Source: *V. Lapparent, M. J. Geller and J. P. Huchra,* Astrophysical Journal, *302, L1, 1986.*

Wavelengths have been stretched by as much as 350 percent. This means that ultraviolet radiation has been transformed into infrared. If a 100-watt light bulb had its light somehow transformed in this way, it would cast no discernible light at all.

As we approach the periphery of our map of the universe, we are examining the way things were much closer to the moment of creation. We are dealing with a universe much younger, smaller, and hotter than today's. When the universe was less than half a million years old, it was so hot that no familiar form of matter could endure. While electrons and atomic nuclei, the constituents of atoms, existed, their motions were too violent to permit the formation of atoms. Under such extremes of temperature, light waves could not propagate freely from one point to another without being absorbed or scattered. The universe was opaque. Suddenly, as the universe expanded and cooled to the point where neutral atoms could be formed, corresponding to a temperature of about 3,000 °C, the universe became transparent. The thin skin of our map of the universe represents the first half million years of its existence, a period of opacity. At the end of this period, the universe was white hot—it was as brilliant as it is at the surface of a star. As the universe suddenly became transparent, it was flooded with a dazzling radiance of supernal brilliance. As we look off into the distance at the night sky, should we not see the light from the primordial Big Bang? Does not the dark of night disprove the theory of a fiery creation?

The answer is simply that the light of the Big Bang does indeed surround us still. Yet, it has been red-shifted so much that it has become invisible to us. According to modern cosmology, the primordial radiation emerging from the hot Big Bang, in its long travels to us today, has been stretched in wavelength by a factor of 1,000. Another way of saying this is that the universe as a whole has expanded and cooled since becoming transparent. At the beginning of the Age of Atoms, it was at a temperature of more than 3,000 °C. In the cold dark universe we inhabit today, the temperature is now only three degrees from absolute zero. The intense radiation produced by the Big Bang is still with us, but it has been transformed into harmless and invisible microwave radiation.

The story of the prediction and discovery of the cosmic background radiation is a curious one. The first experimental evidence for it was obtained in 1937, but no one recognized the significance of the data. George Gamow, in 1948, understood that the early universe had to be very hot in order for it to evolve into the universe of today. His colleagues realized that a remnant of the primordial fireball, in the form of a universal background of microwave radiation, should persist today. However, it seems as if no one took this work as seriously as it deserved.

A modern theoretical approach to the Big Bang universe and its experimentally observable consequences was developed by P.J.E. Peebles and his co-workers at Princeton University. They calculated that the universe should be filled with background microwave radiation which could be found in a dedicated search. The experimental group at Princeton set about to discover this faint residue of the universe's ancient splendor. Before their experiment got underway, they found that they had been serendipitously scooped by two young physicists working at the Bell Telephone Laboratories.

A strange-looking horn antenna was used to discover the cosmic microwave background radiation. It was constructed by Bell Labs in Holmdel, New Jersey for the practical purpose of communicating with our then very primitive satellites. When this device was no longer needed for satellite work, Arno Penzias and Robert W. Wilson modified it for purposes of radio astronomy. They prepared to measure the level of background static in microwaves, an effect which could interfere with observations of distant radio sources. Their expectations were to see no such radio noise at all. Their measurement was intended to produce a negative result, and would merely confirm the quality of their antenna.

An unexpected background signal was apparent in their first experimental run on May 20, 1964. But before they could accept such a result as an indication of a new astrophysical phenomenon, Penzias and Wilson had to exclude more conventional explanations. Had they underestimated the manmade radio pollution coming from the Big Apple nearby? By pointing the antenna directly at New York City, they showed that this was not the case. Or was it caused by something else? "A pair of pigeons was roosting up in the small part of the horn where it enters the warm cab," Wilson recalls. "They had covered the inside with a white material familiar to all city dwellers. We evicted the pigeons and cleaned up their mess, but obtained only a small reduction in antenna temperature." Finally, Penzias and Wilson convinced themselves that they had indeed discovered something new. In 1965 two papers back-to-back appeared in the *Astrophysical Journal*. One was the experimental paper from Bell Laboratories pointing out the inexplicable behavior of the antenna. The other, from the group at Princeton, offered an explanation of the effect employing cosmological theory. In 1978, Penzias and Wilson received the Nobel Prize in physics for their discovery of cosmic background radiation.

Since 1965, many different measurements have been performed on cosmic background radiation. It has been detected at a variety of radio frequencies. Moreover, it has been measured in the infrared region of the electromagnetic spectrum. Since this radiation does not penetrate the atmosphere, these experiments were performed remotely in rockets and balloons. All the data accords with the cosmological interpretation: we are witnessing the last dying whimper of the original Big Bang.

To a very great precision, the background radiation is uniform. This means that the universe, when it just became transparent, was almost entirely featureless and homogeneous. Galaxies had not yet emerged from the primordial chaos. However, the universe does appear to be a very little bit hotter in the general direction of Virgo. This is interpreted in terms of the absolute motion of our Solar System in the universe. Our entire Milky Way appears to be hurtling off towards the Virgo cluster with a breakneck speed of 400 kilometers per second. This is not an entirely unexpected result, since our galaxy is affected by the gravitational forces of neighboring galaxies. It is merely the last vestige of anthropocentrism. Far from being at the center of the universe, we live on a minor planet of a typical, if somewhat small, star. The Solar System lies in the boondocks of a more-or-less average galaxy among billions, a member of an entirely unremarkable "poor and irregular" cluster of galaxies. The whole shebang is not even at rest in the universe. Its motion is controlled by a rather magnificent cluster of galaxies almost 100 million light years away. We may be very special in our own eyes, but we certainly do not inhabit a very privileged position in the universe.

Are We Alone in the Universe?

There are, in the visible universe, a hundred billion galaxies each with a hundred billion stars. Is it reasonable to suppose that life is unique to one obscure planet? The laws of physics are one and the same everywhere. Perhaps there are zillions of other ''earths,'' just as hospitable as ours, teaming with exotic plants and animals. Evolution, having once led to sentience, could have done so again and again. Other societies upon distant planets may also search the skies with wonder, struggle to produce a theory of Nature and seek an understanding of their role within it.

Thoughts like these are not novel, and have recurred over the centuries.

Metrodorus, in the fourth century B.C., suggested that ''. . . to consider the Earth the only populated world in infinite space is as absurd as to assert that in an entire field sown with millet only one seed will sprout.''

In the first century B.C., Lucretius recognized that ''. . . we must realize that there are other worlds in other parts of the universe, with races of different men and different animals.

And others, too, imagined life on other planets and in distant worlds. Giordano Bruno in the sixteenth century, Christian Huyghens, in the seventeenth century and so on until our time. In 1959, Giuseppe Cocconi and Philip Morrison wrote this:

> . . . Near some star rather like the Sun there are civilizations with scientific interests and with technical possibilities much greater than those now available to us. To the beings of such a society, our Sun must appear as a likely site for the evolution of a new society. It is highly probable that for a long time they will have been expecting the development of science near the

> Sun. We shall assume that long ago they established a channel of communication that would one day become known to us, and that they look forward patiently to the answering signals from the Sun which would make known to them that a new society had entered the community of intelligence . . .

Why should we listen for a signal from an extraterrestrial civilization, rather than attempt to send a message to them? The answer is simple: We have the technical competence to listen, but not to send. We Earthlings are a very young species. Like little children, we ought to listen, but not speak. Life has existed upon Earth for the major part of its four-billion year history, and we are merely the most recent of its many tenants. Brute dinosaurs owned the planet for some 300 million years, while something resembling us has been around for less than a million. Recorded history goes back perhaps 10,000 years, but we have had the skills necessary to listen to our brother extraterrestrials for merely two decades.

Is human history near its very beginning, or fast approaching the end? In view of our apparently insatiable appetites for weaponry and war, the pessimist may believe that intelligence is hazardous to its own health and survival. Humankind, or its equivalent elsewhere, annihilates itself in a nuclear holocaust just as soon as it discovers how. I am not such a pessimist. I believe that we shall emerge from the present nuclear danger to flourish anew, just as a greater Western Society sprung from the dark millenium. Our planet is bountiful and can serve as well for a billion years. Today's society is more just, more fair, and more humane than it ever was. Tomorrow will see "life, liberty, and the pursuit of truth and beauty," as well as "from each according to one's ability, to each according to one's desires." To the optimist, human society is at its very beginning, immature, but full of youthful promise.

Assume the optimist's view prevails in at least some developed, extraterrestrial societies. Think of the remarkable technological achievements of just this century—less than a millionth part of the natural lifetime of a sentient species: instantaneous and worldwide radio, television and telephone communication, automobiles and jet planes, earphone stereos for scuba divers, throwaway watches of a precision undreamt of a century ago, the eradication of smallpox and the alleviation of gout, pocket calculators more powerful than a thousand accountants, a real man on the moon, and so on and on and on. Think of what our society could do in another century, or in a million years. Most of our hypothetical aliens will have far surpassed us, having been accumulating knowledge and skills for countless centuries. If they have anything like the curiosity of our own species—and surely curiosity must be a part of intelligence—they are seeking to contact us.

If there are intelligent critters out there, why don't they simply pay us a visit, instead of waiting patiently for us to pick up the telephone? Unfortunately, space travel is easier said than done, even for our more experienced elder aliens. Martians, if there were Martians, could indeed have dropped in. Unfortunately, there are no intelligent species living in our solar system, other than ourselves. There doesn't seem to be any indication of life on any of the planets.

Not so long ago, respected astronomers contemplated the presence of an advanced society upon Mars. Percival Lowell, who had successfully deduced the existence and orbit of Pluto, thought there might be life on Mars. Many observers of Mars had convinced themselves that an intricate network of artificial canals had been engineered on its surface. Lowell's obseratory at Flagstaff, Arizona had been built primarily to better study the planet. But we now know that the canals were an optical illusion. U.S. spacecraft have now landed upon the surface, photographed it, and searched for life. Mars is cold. Its carbon dioxide atmosphere is very tenuous, and there is no sign of life.

Our Soviet colleagues investigated Venus with an unmanned landing and have reported that there is no high society there either: Venus is fiery hot and is surrounded by an atmosphere of highly compressed carbon dioxide. In all likelihood, Earth is the only body in our solar system that has nurtured living things and it is most certainly the unique seat of learning.

Smart aliens, if they exist, live on hypothetical planets of distant stars, at least a few light years down the road. Elementary calculations, based upon known laws, show that it is infinitely cheaper for such beings to attempt to communicate, rather than pay a visit. In absolute ignorance of extraterrestrial economics, we may define "cheapness" in terms of energy requirements. Paul Horowitz, one of the leaders in the search for extraterrestrial intelligence, argues: "A round trip to the nearest star, conducted at 70 percent of the speed of light, would use an amount of energy equal to the total accumulated electric power consumption of the United States for a half million years, and this only if we postulate an ideally efficient rocket powered by a pure matter-antimatter engine! On the other hand, *communication* over interstellar distances via radio transmissions turns out to be not only feasible, but positively cheap. Using current technology, we could communicate with another civilization like ourselves situated on any of the several million nearest stars; an interstellar telegram to one of the farthest of those stars (at a distance of a thousand light years) would cost $1 per word.'' Note that the energy cost of an alien visit is based upon an entirely hypothetical (and doubtful) superior technology, while the cost of communication is based upon the existing electronic capacity of infant Earthlings. We need not fear an invasion of brutish bug-eyed monsters seeking humans as house pets or table delicacies, nor of avuncular prometheans who offer gems and drugs and guns in return for our land. Among our extraterrestrial brothers, talk is cheap, travel prohibitive.

Before setting about to search for extraterrestrial intelligence, we might try to convince ourselves that it is at least possible that life is a commonplace occurrence in the Milky Way. Let me stress that I shall deal with life as we know it: life built upon the chemistry of carbon, dependent upon plentiful liquid water. In so doing, we exclude the possible existence of truly exotic life forms, such as purely hypothetical beasties who adore sulfuric acid, have a selenium based metabolism, or consist of intricate electromagnetic configurations within a hot ionized plasma. This exclusion is without prejudice. Weird forms of life may exist somewhere, and may be attempting to communicate with us as well. May the Force be with them!

Familiar life forms require a familiar environment. They must inhabit Earth-like planets, orbiting a star not unlike our Sun. Stars much smaller than the Sun simply do not produce enough energy to support a habitable planet. Stars much larger than the sun do not live long enough for life to evolve under stable conditions. Double stars seem unlikely sites for the development of life, since planetary orbits would be irregular and seasonal changes of climate too extreme. The only good stars are G-type, single stars like the Sun. Of the 200 billion brilliant stars in the Galaxy, only about 8 billion satisfy these criteria. Notice that I do not consider the possible existence of life in other galaxies. Extragalactic communication would be difficult at best, since other galaxies are millions of light years away. Power requirements for extragalactic transmissions would be enormous, and the time delay between their ''Hi!'' and our ''How are you?'' being millions of years would be far from conducive to productive conversation. Perhaps the time will come for some very advanced and long-lived society to search for messages from other galaxies. For us, it is enough to listen to the words of wisdom from our million nearest neighbors in our own Milky Way.

A sun without planets is like a day without sunshine. How likely is it for a Sun-like star to be surrounded by a system of planets? A system including at least one planet sufficiently like Earth to permit the accumulation and retention of liquid water and the development of life? Astronomers do not yet have a definitive answer to this question. They are not yet able to detect the existence of planets orbiting a distant star. And yet, most believe that planet formation is a common occurrence. Very recently, this notion has been strongly supported by experimental evidence. IRAS is an acronym for the Infrared Astronomy Satellite, a device which was launched into orbit in January 1983 and functioned perfectly for 10 months, until it exhausted its supply of coolant. It was a joint venture of the United States, the Netherlands, and Great Britain. IRAS consists of a 23-inch infrared telescope positioned in an orbit 560 miles above the Earth. Infrared radiation is characterized by wavelengths longer than visible light and shorter than microwaves, typically from 10^{-2} to 10^{-4} cm. Except in narrow windows, infrared light cannot penetrate the Earth's atmosphere and ground-based infrared astronomy is difficult and limited. The data relayed back to Earth from IRAS is still under careful scrutiny, but the preliminary results are extraordinary. For example, it has already expanded the catalog of known infrared sources in the cosmos from a few thousand to a quarter of a million. It has discovered the existence of a curious new class of galaxies that produce more heat than light. One example, known as Arp 220 and located 300 million light years away, emits 99 percent of its energy in the form of infrared radiation. It is believed that these peculiar and unexpected entities are the result of collisions between galaxies, collisions which create a veritable population explosion of new stars.

Perhaps the most remarkable of IRAS' discoveries is the observation of vast shells, or disks, of particulate matter surrounding two young stars, Vega and Fomalhaut. It is from such a cloud of material that the planets in our solar system were formed some five billion years ago. Since the IRAS announcement, similar protoplanetary disks have been observed circling another half-dozen stars. Unfortunately, infrared observations can detect vast clouds of small particles, but not planets that such clouds are believed to become in a few million years. The data suggest that protoplanetary disks are a commonplace, perhaps universal accompaniment of star formation. It is not unreasonable to assume that those clouds condense and coalesce into planets. The betting is that of the eight billion good stars in the Galaxy, most are surrounded by planets.

Not every planet is a good planet. Of the nine in the Solar System, only one is convincingly alive. The initiation and evolution of life on a planet requires a reasonable degree of continuity and a period of billions of years. Some scientists argue that if Earth were only 5 percent closer to the Sun, it would have been subject to a runaway greenhouse effect at the age of a billion years, well before intelligent life could have evolved. Similarly, if it were located only slightly further from the Sun, runaway glaciation would occur as soon as its atmosphere became oxygen rich. It takes great good luck for a planet to avoid the fates of our nearby neighbors Venus and Mars—one too hot, the other too cold. It is estimated that only 5 percent of the good suns have planets which, like Earth, are just right. We are down to a mere 400 million habitable extra-solar planets within the Galaxy. Incidentally, this conclusion is in surprising agreement with the deductions of one Thomas Wright, the astronomer upon whose discoveries Immanuel Kant fashioned his theory of "island universes." In 1750, Wright (on the basis of less than compelling evidence) announced the existence of 170 million habitable planets within the Milky Way.

A habitable planet is not an inhabited planet. What are the chances that a good planet of a good sun will have nurtured life? This is a question we cannot really answer. We know of only one habitable planet, and it indeed has life. This is not an argument suggesting that all nice planets breed life; for if Earth had chosen not to, we would not be here to be the wiser.

What is life as we know it? It is based upon self-replicating DNA—a twisted, complex, organic molecule—the quintessence of all forms of life on Earth. Could such a structure have arisen spontaneously upon a youthful Earth?

When it first condensed from a disk of icy specks and rocky granules left over from the creation of the Sun, Earth was barren—no atmosphere, no oceans, no mountains, no continents. It was a featureless ball, mostly rock on the surface, iron within. Inside the rocky portion, a tiny amount—a few parts per million—was the naturally radioactive elements thorium and uranium. These atoms are subject to a process of spontaneous decay by which they are transmuted into lead with "half-lives" of fourteen and four billion years, respectively. (The "half-life" of a radioactive material is the time period in which half of a given sample of radioactive atoms are subject to decay.) In the process of decay, heat is released. Today, the total amount of heat generated by natural radioactive sources within the Earth's crust is about 3×10^{13} watts (30 terawatts), a hundred times greater than world production of electrical power. In the younger Earth, it was much greater. Without radioactivity, there could have been no life on Earth.

Incidentally, natural radioactivity has played a key role in allowing us to estimate the age of the Earth and the Solar System. Uranium atoms decay into a specific isotope of lead, ^{206}Pb, whereas other isotopes of lead, like ^{204}Pb, are not the daughters of radioactive atoms. Thus, the ratio of the number of ^{206}Pb atoms to the number of ^{204}Pb atoms in a given sample of ore depends upon its uranium content and upon its age or how long the uranium has been decaying into lead. Precise measurements of this ratio have been performed on a variety of rocks. Terrestrial rocks have been discovered with ages of up to 3.8 billion years. The Earth is thought to be older than that, but rocks from its first turbulent phase apparently were never formed or have not survived. Strangely enough, the first fossil indications of life on Earth date back to those early rocks. However, lunar samples are measured to be 4.2 billion years old, and meteorites are established to be 4.6 billion years old. Since it is believed that all of the Solar System was formed at the same time over the span of a million years or so, it follows that the age of the Earth must be 4.6 billion years, plus or minus a hundred million years or so.

The intense heating of the interior of the Earth by radioactive decay led to instability. Trapped within the Earth because of its low conductivity, radioactive heat led to melting, to chemical transformations, and to the development of enormous pressures. It was as if the Earth had developed a severe case of acne. The poisons and pressures were expelled by means of pimples upon the face of the Earth: It was the Age of Volcanoes.

Vulcanism is presumed to be the origin of Earth's primordial atmosphere. It was a very different atmosphere from today's, being constituted by volcanic gases with no life-giving oxygen at all: it consisted of nitrogen, carbon dioxide, hydrogen, water vapor, argon, and a bit of sulfur dioxide to produce a more convincingly hellish ambience. Volcanoes everywhere, continuous rain and thunderstorms, and a lethal atmosphere: not a very promising beginning. Indeed, things were even worse. With no oxygen, there was no ozone layer, and no protection from the destructive ultraviolet rays of the Sun. It was not at all a nice place for birds, bugs, or people.

Of the volcanic atmosphere of Earth, only the argon and some of the nitrogen remain today. The argon is a clue to the early atmosphere: it is not the isotope of argon that is abundant in the Sun and elsewhere in the cosmos. It is an isotope produced by the radioactive decay of a naturally occurring isotope of potassium. Evidently, it was produced by the volcanic outgassing of the Earth's crust, and could not have been there from the very beginning.

Clouds of water vapor developed, just as they do today, and rain fell for millions of years. The waters collected in the valleys about the newly formed volcanoes. The seas and mountains of Earth were formed. Meanwhile, the Sun's ultraviolet radiation worked its magic upon the atmosphere, producing simple, basic gases like methane and ammonia, precursors to the more complex organic compounds that are themselves precursers to life.

What happened next is subject to experimental test in the laboratory. In 1953, Harold Urey and Stanley L. Miller, working at the University of Chicago, concocted a gaseous mixture resembling the primitive terrestrial atmosphere, a mixture of hydrogen, ammonia, methane, and water vapor. Electrical discharges were produced in the gas to simulate the effects of ultraviolet radiation. They observed the copious production of amino acids, organic chemicals which are the universal constituents of proteins, the chemical building blocks of all forms of life. Extrapolating from the laboratory to the Earth as a whole, the Miller-Urey reactions would produce about a ton of organic material per second upon the surface of the primitive Earth, all of which would be collected and conserved within the oceans. In a mere 200 million years, the ocean would have become a thin but nutritious soup containing the equivalent of one bouillon cube of complex organic material per gallon.

The next steps in the evolution of life are not at all clear. Within the ocean, or within the more protected environment of shallow tidal pools, simple organic molecules polymerized into larger and more complex structures, and ultimately into self-replicating systems. We do not know just how this came about. What we do know is that by the time the Earth was 800 million years old, the oceans were teeming with primitive forms of life, forms of algae and bacteria, not so different from some still surviving today. The oldest rocks found on Earth reveal fossils of early life that thrived 3.8 billion years ago.

The metabolism of primitive life depended, as it does today, upon a ready source of energy. Photosynthesis evolved as an efficient procedure to make use of solar energy. Carbon dioxide of the atmosphere was converted into a poisonous by-product which permeated the still sterile atmosphere: oxygen. At first, oxygen reacted with ammonia and methane, until, about two billion years ago, the atmosphere was purged of these unpleasant gases. They had served their purpose well, and were no longer needed. Thanks to our ancestral microbes, the Earth was slowly and steadily being prepared for its eventual inhabitants.

At first, the rate at which living creatures evolved was very slow. Reproduction took place simply and asexually by cellular division. They multiply by dividing. Occasional mutations, caused by cosmic rays or chemical errors, led to new varieties, but these occurrences were rare and more often lethal than helpful. Then, about a billion and a half years ago, sex was discovered. New generations were no longer mere clones of their parents, but shared the genetic complement of two individuals. The rate of evolution hastened, and complex multicellular organisms evolved. As the oxygen content of the atmosphere increased, and as ancient glaciers melted, horrid creatures crawled up upon the land. Not long afterwards, a mere 300 million years, mankind emerged as the dominant species on Earth. After a few thousand years of civilization, and its countless wars and famines and moments of inspiration, our civilization is finally prepared to search the skies to discover whether or not someone or something elsewhere has gotten there first.

What are the chances that, upon the 400 million habitable worlds of the Milky Way, the miracle of life has taken place? What are the chances that life has begotten intelligence, that advanced technological societies have developed elsewhere, and that they are actively seeking to communicate with us now? These are tough questions. Even on a good planet of a good star, there are a number of bottlenecks to the spontaneous appearance of life. Volcanoes resulting from radioactive heating of the crust played an essential role in outgassing the Earth's crust to produce its first atmosphere. The abundance of uranium and thorium varies greatly in the Galaxy, for these elements are created exclusively by supernova explosions. Were other solar systems formed sufficiently near to an ancient supernova so that they too could have become radioactive enough to generate a home for life? Some scientists argue that tidal pools were the creche of primitive life. The tides depend upon the Moon. No other planet in the Solar System has a moon as beautiful and as influential as our own. Must other habitable planets have large and nearby satellites in order for life to emerge? Are continents a universal feature of habitable planets, or could intelligent creatures evolve at sea? How could a beast of the sea, however brilliant, discover the stars and begin, by wondering, upon the long road of scientific discovery? Questions, questions, questions. Let us say, as many of my distinguished colleagues have done, that 10 percent of the good planets of good suns generate life. And, let us assume that life inevitably evolves to a level of intelligence, creativity, curiosity, and civilization at least as great as our own. With this estimate, we still count some 40 million inhabited planets in the Galaxy, in each of which life has evolved and intelligent society has developed. But, how long does a civilized society survive? In order for a distant race to communicate with puny Earthlings, it must be a patient and long-lived society. Humanoids have walked the Earth for a million years, but we have only just convinced Senator Proxmire that the search for extraterrestrial intelligence is a sensible and serious scientific enterprise. Our hypothetical alien civilization, if it is to succeed in contacting us, must solve its own social and economic problems so well that its advanced and enterprising technology can survive for a million years. Is this at all likely? Let us look to our own society for guidance.

Recorded human history goes back only some ten thousand years. However, we have only just begun to search for messages from outer space. The Roman Empire and the British Empire each lasted for only a few hundred years. The promised millenium of the Third Reich survived for less than a decade. How long can we expect today's advanced technological society to persist? There are a number of clear threats to the indefinite continuation of our species. Some of them, now foci of political attention (and sometimes, of malign neglect), are short-term problems created by our own species' shortsightedness.

Nuclear Holocaust

Only within the past few decades has human society possessed the ability to destroy itself—if not quite utterly then utterly enough. The nuclear armament of the superpowers is sufficient to bestow upon each blessed human being the equivalent of 10 tons of TNT. Yet, the arms race is being pursued furiously by both sides and only in the last days have we seen signs of some lessening of tensions. Their SS20s must be compensated by European Pershings, and our MX simply must be put somewhere. Thirty thousand tanks are deployed in East Germany as a demonstration of the peace-loving nature of the Soviet government. Strange and unpleasant goings on in Afghanistan and El Salvador do not convince us that the leaders of the superpowers are devoted to world peace, security, and prosperity. Nor is much of the activity elsewhere on Earth very reassuring: not the interminable and uncomprehensible war between

Iraq and Iran, not the repression in South Africa, nor the corruption in Zaire, nor the economic disasters of South America and Poland. The world is a dangerous place indeed, and the weapons already exist to destroy us all. And more are being built. Yet, the Soviet and American governments, at odds for 40 years, have successfully avoided any direct military confrontation, let alone a full scale war. They have signed and abided by treaties limiting nuclear tests, banning nuclear weapons in space, and controlling antiballistic weapons development. One day, perhaps, they will succeed in reducing and all but eliminating their stockpiles of nuclear weapons. Let it be sooner rather than later.

Pollution and Erosion

Our bacterial cousins labored a billion years to give us pure air and clean blue skies. Denizens of Mexico City and Los Angeles realize just how effectively we can undo their great enterprise. Fortunately, the problem of smog is both controllable and limited. Emission control devices and more efficient automobile engines work well. Furthermore, we will most certainly begin to run out of petroleum in a century or two, and automobiles as we know them will soon become extinct. Coal is a more serious problem. We, in this country (unlike France, Taiwan, and a few others), have rejected nuclear power as a source of electricity. The blame for this unfortunate action is widely distributed over industrial incompetence, governmental bungling, and public misconceptions. Be that as it may, we are probably primarily dependent upon coal for decades, if not centuries. (We do have enough coal for a few centuries, if we are willing to undergo the strip-mines' sacrifice of several pretty but sparsely settled states.) Burning coal, unless it is done under prohibitively expensive circumstances, produces acid rain. Acid rain, while probably not disastrous to your health, is lethal to trees. The German Black Forest is dying, as are many of the gorgeous Swiss alpine forests. The trees are not yet dead, but they show signs of irreversible damage. The foliage season is still glorious in New England, but for how much longer? Perhaps we really do not need the forests; some of our government officials, after all, seem to regard them as mere paper factories, and who needs paper in the age of computer diskettes? But if Maine loses its forests to acid rain, how long will it be before Iowa loses its wheat, and our Russian customers are forced to revert to cabbage and parsnips? And, for you gourmets, don't believe for a minute that Minnesota wild rice is immune to pollution of air and water.

Erosion goes hand in hand with pollution. Remove the trees, and the soil underneath will be irretrieviably lost. Sicily, once the breadbasket of the Roman Empire, was reduced to a minor producer of olives and grapes by generations of shipbuilders who deforested the land. Certain Arab societies, regarding a barren tree as worthless and sinful, succeeded in reducing much of the once fertile Middle East to desert. And so it continues today in sub-Saharan Africa and in much of China, and in the fertile plains of America. Clear air, pure water, and good soil are essential to efficient agriculture. Without them, no amount of synthetic fertilizer will suffice. How will things look a thousand, or a million years hence? Are we prepared to stay on Earth as long as it will have us, or will we soon make our entire planet unfit for human habitation?

Reproduction

Reproduction, not birds and bees, but babies. At its present rate of growth, the population of the world doubles every 40 years or so (Figure 1). This is a relatively new trend which began two or three centuries ago. It led Thomas Robert Malthus to conclude in 1798 that the rate of growth of population necessarily exceeds the rate of growth of production, so that poverty is mankind's inescapable fate. In large measure, despite the miracle of the green revolution and the modernization of world agriculture, Malthus has been proven right. It is probably true today that three-quarters of the world's population is hungry three-quarters of the time. Clearly, the curve of population growth cannot continue to increase as it has for more than another century or so. One way or another, it will level off. I doubt whether any agricultural advances can feed a population of 25 billion (in 2100 A.D.) or of more than 100 billion (in 2200 A.D.). Something has got to give. One possibility is that the world population will stabilize by controlling its birth rate. Population control can be established by harsh law (as in China), by sterilization, by abstinence, by contraceptive devices, by the pill, by abortion, by infanticide, or by sexual preference. Each procedure has its advocates, and each has its ardent foes. A second, laissez faire, scenario would let Nature take its course. As the population of the world begins to exceed its natural limit, an increased death rate will compensate for a high birth rate. Hunger, misery, and a short, hard life will be the lot of

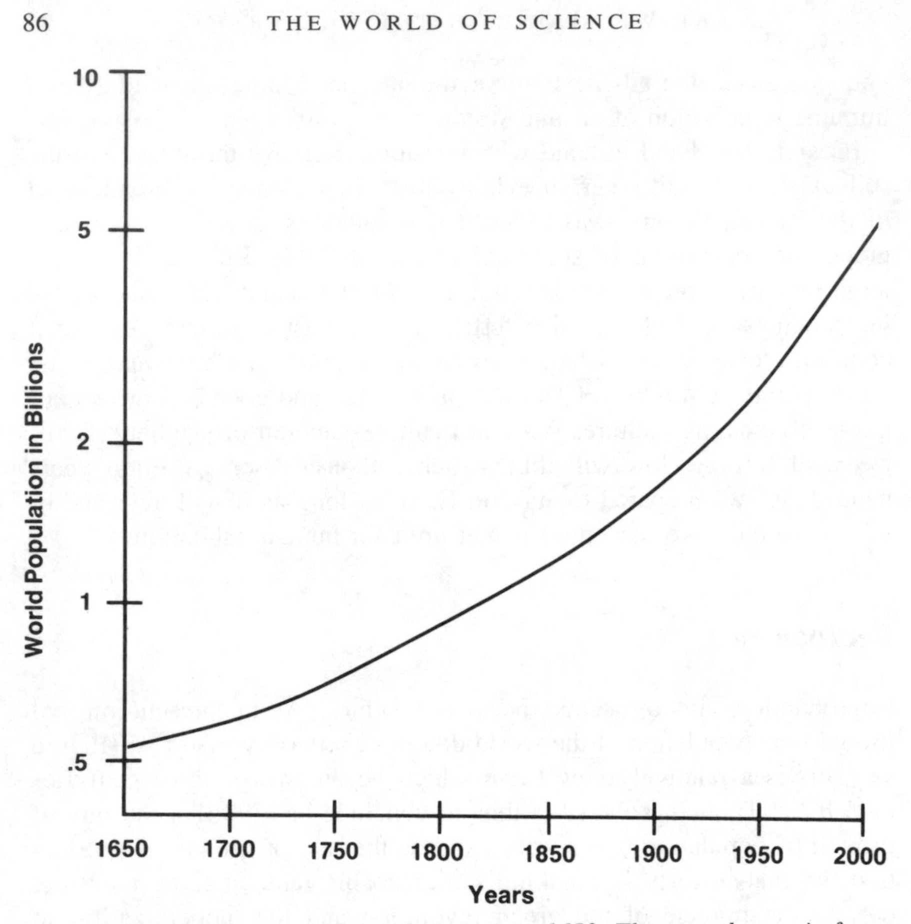

FIGURE 1. *Growth in world population since 1650. The growth rate is faster than exponential.*

the common man—hardly a situation in which a civilized society can be maintained. Population control seems to be an absolute prerequisite for the establishment of a truly long-lived society. Brute dinosaurs thrived off the Earth for 300 million years. If Homo sapiens intend to enter the longevity sweepstakes, men and women had better learn to control their bodily functions a little better.

Exhaustion of Natural Resources

Society depends upon fossil fuels for much of its energy. Coal, oil, and natural gas exist in limited quantities. At the present rate of consumption, in a few more centuries nothing will be left of our fossil heritage. True, we will very likely have developed alternative energy strategies: safe nuclear fusion and solar power, as well as reduced per capita consumption. However, petroleum happens to be a unique and irreplaceable source of a variety of industrially important chemicals. Squandering our petroleum simply for its energy content is a hideous crime that we visit upon future generations. A technological society without a petrochemical industry is conceivable, but far more difficult that it would have been otherwise. The story of natural gas is similar. Natural gas, consisting of a mixture of rather simple chemical gases, is not particularly important to the chemical industry. It burns cleanly, and may as well be burnt now to satisfy current energy demands. However, nature has seen to it that natural gas is our unique natural reservoir of helium gas. Helium is certain to become an essential element in the technology of the next century. It is an ideal and irreplaceable coolant for superconducting devices for efficient energy transmission and storage. Moreover, ^{3}He, a rare isotope of helium, is likely to be the fuel source of early fusion-based power generators. A wise society would extract the precious helium from natural gas (an inexpensive procedure) and put it back into the ground to await future needs. What is done in fact is to extract just enough helium as can be sold—mostly for children's balloons and the two extant Goodyear blimps. The vast majority of our helium, what may soon be needed desperately to resolve a *real* energy crisis, is simply vented into the atmosphere where it is lost forever. Petroleum and helium are but two examples of nonrenewable resources now being recklessly wasted. There are several other scarce resources (like chromium) which are sure to be in very short supply in the not-so-distant future. Unless we change our ways, we shall earn and deserve the curse of our descendants.

It may seem unlikely that human society will choose to survive these threats: Nuclear war, pollution and erosion, population explosion, and the exhaustion of resources. Trends today suggest that we may, at best, anticipate a century of future progress, not a millenium, not a million years, not a billion years. After that, one or another of the problems will get out of hand and we will begin our slow regression towards something less than a humane civilization of promise and hope. Each successive generation may look forward to a somewhat tougher, meaner, and shorter life. Parents may describe to their unbelieving offspring the lost wonders of the world, like television, washing machines, and breakfast cereals, that they knew when they were children. The age of downward mobility may begin. Perhaps it has already. A civilization more clever than wise, more fertile of body than of mind, more concerned with the pursuit of happiness than the fulfillment of promise can only decline.

Perhaps we shall avoid our seemingly inevitable fate. Things can change. The Dark Ages were followed by the Renaissance. Many of our citizens are becoming aware of the fragility of the environment and are acting upon their concerns. All of these problems were created by human beings. Miracles are possible in the sphere of human accomplishment. As far as our hypothetical aliens are concerned, let us imagine that they will have overcome their social obstacles and have built stable, long-lived, and inquisitive societies.

I shall assume, to pick a number out of a hat, that the average survival time of an alien society is five million years: small compared to the tenure of dinosaurs on Earth, and large compared to the age of primates. A survival time of five million years assumes that a civilization can deal with natural catastrophe as well as unintended racial suicide. On Earth, for example, the ice age will return in 20,000 years or so. Most of the United States and Europe will be buried under glaciers hundreds of feet thick. Not a very serious problem for a truly advanced society: we simply move south. A more serious event will take place 13 million years from now. It appears as if Earth is subject to a devastating barrage of comets every 26 million years or so. The last two occurrences were relatively mild, but the barrage of 6.5×10^7 B.C. led to a nuclear-winter-like scenario in which the Sun was obscured for several years. It annihilated the dinosaurs. Whether or not our society will survive the next recurrence is impossible to determine, but we do have a long time to prepare for it.

Let us summarize our speculations about extraterrestrial intelligence. There are 200 billion stars in the Milky Way, of which some 8 billion closely resemble our sun. All, or almost all, are likely to be accompanied by planets. Four hundred million of these solar systems contain planets much like the Earth, and upon 40 million of these planets life has emerged, evolved, and flowered into an advanced technological civilization. However, the estimated lifetime of an advanced society is only five million years—a mere thousandth part of the star. A distant society must be alive and well in order to send us a telegram, and according to our estimate there are only 40,000 such societies in existence in our galaxy right now. Taking into account the density of stars in the neighborhood of the Sun, we conclude that there may be a *few* extant extraterrestrial societies within a thousand light years of Earth. Perhaps they are in contact with one another. Perhaps they are trying to establish contact with us. How should we go about initiating an extraterrestrial conversation?

Visits in person, as we have seen, are out of the question. We could, in principle, beam messages to other stars. However, there are some 40,000 sunlike stars within a thousand light years of Earth, and to direct antennas at each would be prohibitively expensive at the present stage of human technology. We have just achieved a technological society, whereas they have been at it for hundreds of millenia. We, to them, are as children, and it is for the child to listen, not to speak.

The first modern and serious scientific approach to the search for interstellar communications was pioneered by Giuseppe Cocconi and Philip Morrison in 1959. In their words ". . . It follows, then, that near some star rather like the Sun there are civilizations with scientific interests and with technical possibilities much greater than those now available to us. To the beings of such a society, our Sun must appear as a likely site for the evolution of a new society. It is highly probable that for a long time they will have been expecting the development of science near the Sun. We shall assume that long ago they established a channel of communication that would one day become known to us, and that they look forward patiently to the answering signals from the Sun which would make known to them that a new society had entered the community of intelligence. What sort of channel would it be?"

There is little doubt that the only practicable method of interstellar communication involves transmission of electromagnetic waves. Since the advanced alien is attempting to communicate with a newly evolved (developing) society, it will probably attempt to make reception as easy as possible. It will choose a broadcast channel that passes through the Earth's atmosphere and can be received by ground-based listeners. The extraterrestrial signal must lie within one of the two electromagnetic windows of the atmosphere: visible light or radio waves. The visible option seems untenable. The power requirements to send a recognizable light signal over distances of hundreds of light years are enormous and it would be a formidable task for us to disentangle the signal from the enormous background of light coming from an alien sun. Communications will be established by radio signals, with wavelengths lying within the atmospheric window, from one centimeter to thirty meters.

In 1967 some scientists believed that they had detected an interstellar radio beacon from a communicative alien society. Jocelyn Bell Burnell, then a graduate student at Cambridge University, observed a very rapid and regular series of radio pulses coming from a point source beyond the Solar System. The pulses were very regularly spaced in time, arriving just 1.3373011 seconds apart. For a time, the source was called a LGM, for Little Green Man, for what besides an alien society could fashion a clock of such perfection? (Why it was called a Little Green Man and not a Large Lilac Lady remains a sexist mystery.) Soon, however, Burnell discovered other LGMs elsewhere in the sky, with pulse periods of .253065, 1.187911, and 1.2737635 seconds. It became clear that these were not Little Green Men at all, but that they represented a new class of natural astronomical objects. They are now known as pulsars and are believed to consist of rapidly rotating neutron stars, relics of recent supernovae. In the search for extraterrestrial intelligence, it was just a false alarm.

To what wavelength should we tune our radio receivers in order to detect a message from the stars? Edward Purcell, my colleague at Harvard and a Nobel Laureate in physics, puts the question very well,

> If you want to transmit to a fellow and you can't agree on a frequency, it's nearly hopeless. To search the entire radio spectrum for a feeble signal entails a vast, and calculable, waste of time. It is like trying to meet someone in New York when you have been unable to communicate and agree on a meeting place. Still, you know you want to meet him and he wants to meet you. There are only two or three places: Grand Central Station, among other key places. Here, there is only one Grand Central Station, namely, the 1,420 megacycle line which is, by a factor of a thousand at least and probably more, the most prominent radio frequency in the whole galaxy. There is no question about where you transmit if you want the other fellow to hear, you pick out the frequency he knows. Conversely, he will pick out the frequency he knows we know, and that is the frequency to listen to. If you play this game carefully, you will find the conclusions inescapable. We know what to do; we know where to listen.

What is this 1,420 megacycle (or, in terms of wavelength 21 cm) radiation to which Purcell alludes? It is a frequency uniquely characteristic of a hydrogen atom. Crudely oversimplifying, we may regard the hydrogen atom as a planetary system in which one electron revolves about a central proton. Each constituent particle is spinning like a top, and each particle affects the other. As a result, the two spins are in a state of precession, turning about one another at the rate of 1,420,405,751.768 rotations per second, or, at a rate of roughly 1,420 megacycles. Notice that this number is known to the ludicrous precision of 13 decimal places. Known as the hydrogen hyperfine frequency, it is one of the best measured physical quantities. Its measured value is in absolute agreement with its value as computed from fundamental theory. This is one reason why today's physicist is jubilantly and justifiably confident of his understanding of the structure of the atom.

Because the natural frequency of vibration of a hydrogen atom lies at 1,420 megacycles, it turns out that hydrogen atoms in space should emit electromagnetic radiation at just this frequency. Since hydrogen is by far the most plentiful chemical element in the cosmos, it is not surprising that its characteristic 21-cm electromagnetic radiation is the unique and dominant feature in all of radio astronomy. That the 21-cm radio emission line from space could be detected by radio astronomy was pointed out by H.C. van de Hulst in Holland in 1944. The 21-cm line was first observed by none other than Professor Edward Purcell, working at Harvard with his graduate student, Harold Ewen, in 1951. It marked the beginning of the important discipline of spectral-line radio astronomy. Despite Purcell's parietal interest in 21-cm radiation, his argument that it (or one of a small number of nearby and related magic wave lengths) is the obvious choice for interstellar communication is nonetheless compelling.

The first serious search for extraterrestrial intelligence (SETI) was carried out in the early sixties by Frank Drake in a pioneering experiment known as Project Ozma. It was directed at a few nearby stars, and not surprisingly, no signal was found. Much more ambitious (though not very expensive) searches are being carried out today. For example, a SETI is being conducted out now by my Harvard colleague, Paul Horowitz, at the modest 84-foot Oak Ridge radiotelescope. Each minute of running develops the equivalent sensitivity of a thousand years of the operation of Project Ozma. Ultimately, we shall be able to direct our antenna at all the promising stars within hundreds or thousands of light years from Earth. If there is a beacon, then sooner or later, it will be found.

There is no question that SETI is a risky enterprise. It is impossible to conclude with any degree of certitude whether extraterrestrial intelligence exists or not. Nonetheless, it is an essential pursuit. I agree with Horowitz when he says, "Communication with an extraterrestrial intelligent species would be the greatest single discovery in the history of mankind." Think of the things we might learn from a distant civilization that has come to terms with its environment so as to flourish for a million years. Imagine its art, its music, its history, and its science. Even Alfred

Adler, an irrepressibly rational critic of the SETI program, is forced to conclude: "The reception of a galactic message would be significant most of all because it would replace a human expectation with a certainty. It would prove to humanity that we are not alone in the universe, and in proving might diminish man's self-consciousness and self-centeredness. The consequences could only be beneficial. And that is about all the profit that could be expected." Death, taxes, and *neighbors*! Just to know that we are not alone would be reward enough, for then we would know that we, too, can overcome.

The Big Picture

The most fundamental sciences are those which probe most deeply into the structure of matter, the nature of space and time, the origin, evolution, and fate of the universe. These fields are cosmology and elementary-particle physics. Under different guises, these disciplines have been with us since the very beginning of human civilization. Perhaps the discovery of fire was the first step on the way to modern elementary-particle physics; and the observation of the regular patterns of day and night, summer and winter, led to the first cosmological speculations. We have come a long, long way, and have answered many profound questions en route.

Today, we know beyond any doubt that our universe is expanding. The billions of galaxies that comprise the observable universe are hurtling apart from one another, rather like debris from a titanic explosion. Most distant galaxies and mysterious quasars recede from us at nearly the velocity of light. The initial explosion, in which not only matter, but space-time itself, was born, took place a long, long time ago. We cannot pinpoint its birthday precisely, but we know that it was about 15 billion years ago. We call it "the Big Bang."

Our contemporary universe, considered as a whole, is extremely cold and extremely tenuous. A parcel of space the size of our planet contains, on average, only about a microgram of matter. (Of course, Earth is far from a typical piece of real estate.) The mean temperature in the universe was first measured in the 1960s. It is a mere 3 °C away from absolute zero. (Again, Earth is a special place upon which the temperature is a hundred times greater than the universal mean.) But as our universe expands, it continually becomes thinner and colder. Earlier, it was hotter and denser. When the universe was a mere million years old, its mean temperature was that of the surface of the Sun. At the tender age of about three minutes, its mean density was that of water and its temperature was billions of degrees Celsius. At an earlier moment yet, you, me, our solar system, and every part of the known universe, would fit comfortably on the head of a pin! How did our universe evolve from the moment of the Big Bang until now? How did stars and galaxies develop from the primordial furnace? At what point, and how, were the various chemical elements formed? What will happen to our universe in the remote future? Why are the laws of nature what they are? Why is space three dimensional? What is the ultimate connection between gravity, which determines the structure of the macroworld, and quantum mechanics, ruler of the microworld? These are some of the questions addressed by the modern cosmologist.

We know many of the secrets of the structure of matter. In current theory, there are precisely 17 apparently elementary entities: the building blocks of all known forms of matter and the agents of all observed interparticle forces. Today's "matter particles" can be organized into a form of periodic system, the table of quarks and leptons provided in Table 1.

TABLE 1. Quarks and Leptons

Charge =	*−1*	*−1/3*	*0*	*+2/3*
I	Electron	Down Quark	Electron Neutrino	Up Quark
II	Muon	Strange Quark	Muon Neutrino	Charmed Quark
III	Tau	Bottom Quark	Tau Neutrino	Top Quark

The pattern includes six varieties of quarks, and six varieties of particles, known as leptons. These 12 particles are separated into columns according to their electrical properties or, more specifically, their electric charge. The rows correspond to the three "families" of fundamental matter particles. The first family includes those ingredients that are required to build atoms. Up and down quarks cling to one another in groups of three to form "nucleons." These, in turn, are the basic components of every known atomic nucleus. Negatively charged electrons are attracted to nuclei to form electrically neutral systems: atoms. Atoms link together to form molecules, and vast assemblages of molecules form familiar objects of ordinary life: bugs, birds, bricks, bombs, binary stars.

What about the electron neutrino? Does this last remaining member of the "first family" have a role to play? It is not a part of matter, to be sure. Neutrinos play a central role in one of the three forms of natural radioactivity, the so-called "beta process." It is by means of this process that neutrons and protons can change into one another. In particular, it is an essential mechanism for the fusion of hydrogen into heavier nuclei that takes place in the interior of every star—a process crucial in two ways for life on Earth. The very material of our planet was fabricated in the furnaces of stars long ago. Warming rays of the sun are produced in the sun by nuclear fusion. If there were no electron neutrino, none of this could take place. Spaceship Earth would be a solid ball of frozen hydrogen. Thus, the first family of quarks and leptons, in its entirety, is essential for the existence of the world as we know it.

What is the ultimate purpose of the second and third families? They do not seem to play an essential role in everyday phenomena. Technology has found no use for the muon, nor for the charmed quark. Nonetheless, these curious particles *do* exist. They are produced by cosmic rays and at particle accelerators. They were a dominant component of the universe in its infancy. Why are there three families when one would seem to suffice? This is the problem that particle physicists call the "question of superfluous replication." We have no clear answer yet, but there is one hint:

There are a number of curious asymmetries in nature. One is the failure of "time-reversal invariance," the existence *even in the microworld* of a well-defined arrow of time. In the everyday world, there is an evident difference between the past and the future. People, for example, pass from birth to death and never the other way around. This we understand. It has to do with the tendency of complex systems, with very many parts and pieces to become disordered without active intervention—a fundamental principle known as "the second law of thermodynamics." But atoms and elementary particles are not such complex systems. For them, we once thought, there was no absolute distinction between past and future. We were wrong. A group of experimenters laboring at the Brookhaven National Laboratory discovered that even the most elementary systems can "sense" the arrow of time.

Another of nature's asymmetries has to do with matter and antimatter. Corresponding to each matter particle there is an antiparticle with exactly the same mass, and exactly opposite electrical properties. Matter and antimatter cannot coexist under normal conditions. Earthly things are made of matter exclusively but, in principle, they could have just as well been made of antimatter. Antimatter can be created, in minute quantities, in high-energy physics laboratories. All told, in all the laboratories of the world, not more than a milligram of antimatter has ever been made. Our entire solar system is made of matter. So is the Milky Way and nearby galaxies. Scientists believe that the entire known universe is made of matter, with hardly a bit of antimatter. Why is this? Why is nature biased in favor of antimatter?

Andrei Sakharov, winner of the Nobel Prize for peace in 1975, is known to the world for his courageous defense of human rights in the Soviet Union and elsewhere. It is less well known that he is one of the greatest living physicists.* His early work on nuclear fusion is the basis of modern attempts to harness the mechanism of the solar furnace to produce electricity on earth. In 1967, he saw a connection between the matter-antimatter asymmetry and the time-reversal asymmetry. In the context of theories that had not yet been developed, he showed how the one could produce the other. The effect of time-reversal asymmetry, in the very young and hot universe, was to force it to become matter (rather than antimatter) dominated. Not only did Sakharov point the way to the solution of a profound puzzle, but he was among the first to show how deeply cosmology and particle physics are related—in this case, how a discovery in particle physics can explain how the universe may have evolved.

What does all this have to do with the three families of quarks and leptons? Sakharov's argument is based on today's ''grand unified theory'' in which the violation of time-reversal symmetry is easily accomplished and, indeed, it happens automatically—only if at least three families of fundamental matter particles exist. The extra families may not be as superfluous as they seem. Without them, there is no natural violation of time-reversal symmetry and, consequently, the universe would be balanced equally between matter and antimatter. In such a universe, there is no Sun, no Earth, and no people to wonder at how it works.

Sometimes the connection between particle physics and cosmology is reversed; astronomical observations constrain the properties of the microworld. An astonishing example concerns observations about the abundance of the element helium in the universe. Loosely speaking, three-quarters of the matter in the universe is in the form of hydrogen, with

* *Author's note:* Sakharov died in 1989.

most of the rest, helium. The other elements, such as those that comprise most of the Earth, account for no more than 1 percent of matter in the universe. Most of the helium is primordial; that is to say, it was produced in the hot Big Bang. The rate of helium production in the early universe depends critically on the number of different kinds of neutrinos that exist. Astrophysicists insist that there can exist no more than four neutrino species.

Putting it all together, we see that cosmological arguments demand that there be three or four families of matter particles. Experimental physicists have so far discovered three such families.*

The fundamental structure of matter has two logical components which might be called, being and becoming. So far, we have enumerated 12 varieties of matter particles comprising the aspect of being. These particles must combine with one another in order to form the stuff of the universe. They must interact in order to become higher forms of organized matter; they must act upon one another. The processes of becoming are described by forces. To each distinct variety of force there corresponds a force particle in our modern and powerful mythology. The five known force particles are summarized in Table 2.

TABLE 2. Known Force Particles

Graviton	The agent of gravitational force
Photon	The agent of electromagnetic force
Gluon	The agent of the strong nuclear force
$W^{\pm}$ and Z^0	The agents of the two varieties of weak nuclear force

Of these, two have only just been produced and observed in the laboratory. It was for their efforts towards this great discovery that the physicist Carlo Rubbia and the engineer Simon Van der Meer shared the 1984 Nobel Prize in physics.

The choice of Rubbia and Van der Meer was not a great surprise to the scientific community. These men had been responsible for the design, construction, and deployment of a huge experimental facility at CERN, the European Center for Nuclear Research. It began operating in 1981, and rapidly churned out one great discovery after another: First the W boson, then the Z boson. What was surprising to *me* was to be invited

* *Author's note:* Recent experiments done at CERN and SLAC show that there exist precisely three families of matter particles.

to attend the Nobel ceremonies in December 1984. The Nobel Committee wished to recognize the profound connection between the 1984 award, and the 1979 prize which was shared by Abdus Salam, Steven Weinberg, and myself.

Why had the Nobel Committee taken the unusual action of inviting former laureates? Abdus, Steve, and I had received the prize for our various contributions towards the construction of a unified theory of the weak and electromagnetic interactions. The central predictions of the theory were the existence and the properties of the then hypothetical particles, W and Z. The Swedes had taken a gamble. Suppose our theory had turned out to be wrong? The experiments done at CERN represented the ultimate vindication of our theory and justification for our prize. The predicted particles existed, and they had just the properties we had predicted.

The seventies and early eighties were a time of revolution and reconstruction in the world of elementary-particle physics, a time during which a new and powerful theoretical system evolved. The sixties was a decade of mystery. So many new and puzzling elementary particles were discovered that particle physicists were compelled to carry booklets specifying the names and properties of the denizens of the elementary-particle zoo. Two different philosophies arose.

Geoffrey Chew, of the University of California at Berkeley, was an ardent advocate of "nuclear democracy," the view that each of the particles had an equal claim to being elementary. But the population explosion had produced literally hundreds of such particles, and if each were equally elementary then none could be particularly elementary at all. Murray Gell-Mann, of the California Institute of Technology, took the opposite point of view. He imagined the existence of simpler things out of which elementary particles were somehow to be made. He called his hypothetical constituent particles, "quarks."

It was difficult to choose between the two points of view, since neither was backed up by a specific and realistic theoretical framework.

Today, all of this has changed. We have a new and consistent theory of weak and electromagnetic interactions called, unimaginatively, the electroweak theory. Many experiments of the past few years have served to confirm the validity of this theory: the discovery of neutral currents in 1973 and of W and Z in 1983 in Europe, and the discovery of the J/ψ particle in 1974 and of atomic parity violation in 1978 in the United States. In addition, we have a powerful new theory which purports to explain so-called strong interactions. It explains how quarks bind together to form particles, and how these particles combine together to form atomic nuclei. This theory, called quantum chromodynamics, together with the electroweak theory, form what is known as the "standard theory." I have written of this synthesis that "it offers a complete, correct, and consistent explanation of all observed elementary-particle phenomena." I have gone on to speculate that we have come to the end of the discovery of unexpected new elementary particles or fundamental forces. I have been accused of declaring that the discipline of elementary-particle physics has come to its natural end.

I am confident and hopeful that I am wrong. Totally unexpected new things surely remain to be discovered by ingenious experimenters. Nature cannot have exhausted her bag of tricks. Frequently, in the search for the ultimate constituents of matter and the laws of nature, the view was expressed that we have come to the end of the road. Such was the case, for example, at the culmination of the era of classical physics in the late nineteenth century. The great successes of Newtonian mechanics, of Maxwell's theory of electromagnetism, and of the new science of thermodynamics made it appear that the end was in sight. Then, in rapid succession, a series of unanticipated experimental discoveries pointed to the overthrow of the classical system and to revolutionary new theories of quantum mechanics and relativity.

Again, in the late forties, it appeared to some physicists that we had achieved, or at least approached, total understanding. George Gamow was a great research scientist and an effective popularizer of science. His book, *One, Two, Three, Infinity,* published in 1947, was instrumental in convincing me to become a physicist. In it, he wrote:

> "But is this the end?" you may ask. "What right have we to assume that nucleons, electrons, and neutrinos are really elementary and cannot be subdivided into still smaller constituent parts? Wasn't it assumed only a half century ago that the atoms were indivisible? Yet what a complicated picture they present today!" The answer is that, although there is, of course, no way to predict the future development of the science of matter, we have now much sounder reasons for believing that our elementary particles are actually the basic units and cannot be subdivided further. Whereas allegedly indivisible atoms were known to show a great variety of rather complicated chemical, optical, and other properties, the properties of elementary particles of modern physics are extremely simple; in fact they can be compared in their simplicity to the properties of geometrical points. Also, instead of a rather large number of "indivisible atoms" of classical physics, we are now left with only three essentially different entities: nucleons,

> electrons and neutrinos. And in spite of the greatest desire and effort to reduce everything to its simplest form, one cannot possibly reduce something to nothing. Thus it seems that we have actually hit the bottom in our search for the basic elements from which matter is formed.

Immediately after these words were written, the first of the apparently elementary particles called *mesons* was discovered. Soon afterwards, strange particles were seen. With the development of new experimental techniques and the exploitation of large particle accelerators, the trickle soon became a deluge. Gamow's three elementary particles were increased in number by a hundredfold. Clearly, we had not nearly hit the bottom.

Our search for the ultimate structure of matter is an antique discipline. Some ancient Greek philosophers had identified fire, earth, air and water as the ultimate constituents of matter. It soon became clear that such substances as gold, sulfur, or mercury could not be explained by the four classical elements. Chemistry, or its dark sister, alchemy, remained confused and chaotic disciplines until the time of Lavoisier. It was Lavoisier, more than any other scientist, who gave a precise meaning to the Greek notion of the chemical element.

In the two succeeding centuries, our search for the building blocks of matter followed a remarkably cyclical trajectory. Four times in the history of our science, it appeared as if we had "hit the bottom." And four times we have been undone. Whether or not today we have at last succeeded in identifying the innermost layer of the onion is unclear. But, it is illuminating to consider the history of the search.

Four times we had thought we had identified the most elementary of particles: once they were atoms, then nuclei and electrons, then "elementary particles," and now quarks and leptons. In each case, the search has followed a logically similar path:

A. A new level of the onion is revealed or deduced. Candidates emerge as the ultimate constituents of matter.
B. A theoretical system evolves to understand the particles. It emerges as the "elementary-particle physics" of its day.
C. A "population explosion" of allegedly fundamental particle is uncovered. There seem to be too many for all to be regarded as truly fundamental.
D. Scientists discover a regular pattern among elementary particles.

They are grouped into patterns or tables, and they display systematic variations of chemical or physical properties.

E. Experiments reveal hints of further structure. The allegedly elementary systems display behavior of composite systems made up of even simpler things. When these hypothetical entities are found to exist, compositeness is established, and we proceed again to step A.

Let us continue our synopsis of elementary-particle physics with John Dalton, generally regarded as the founder of the atomic hypothesis. He showed how the laws of chemical combination demanded the existence of atoms. The science of *chemistry* rapidly evolved to describe the interactions among fundamental atoms. As chemical techniques became more sophisticated, more and more atomic species were discovered. Lavoisier's list of 30 chemical elements was doubled by the mid-nineteenth century. Who could believe that nature employed so many different fundamental building blocks of matter?

In 1859, the periodic table of chemical elements was devised. It accurately predicted the properties of elements yet to be discovered. It appeared to many scientists of the time that the days of atoms as truly fundamental systems were numbered. Each type of atom had a variety of intrinsic properties: mass, size, and chemical affinity, for example. Moreover, each chemical element, when heated, produced a characteristic spectrum of emitted light, leading to the powerful analytic tool of atomic spectroscopy. The atom was, in fact, a complex structural entity made up of parts.

The electron was discovered in 1896 and the atomic nucleus in 1911. Rutherford realized that the atom consisted of a number of electrons orbiting about a small, dense central nucleus. We were back to stage A.

The new science of *atomic physics* evolved to explain the structure of atoms. A new and revolutionary methodology, quantum mechanics, supplanted the earlier and merely approximate classical theory. Many properties of bulk matter were explained by the new synthesis. However, the problem of the large number of atomic species was not resolved. Indeed, with the discovery of isotopes, it became clear that there exist many more different atomic nuclei than there are chemical elements. A systematic pattern again became apparent. Each atomic nucleus has a definite value of electric charge. Its charge is always *precisely* an integer multiple of that of the electron. For the hydrogen nucleus, it is unity; for uranium it is 92. This integer is known as "atomic number." There is a second integer associated with each nucleus, its weight. Every atomic nucleus has an approximately integral weight compared to the hydrogen nucleus. For example, the oxygen nucleus weighs about 16 times that of hydrogen, whereas the prevalent isotope of uranium has an atomic weight of about 238.

In the process of radioactive decay, both the atomic number (Z) and the atomic weight (A) change in a regular fashion. There are three processes of radioactive decay. In "alpha decay," A is reduced by 4 while Z is reduced by 2. In the process of "beta decay," A is unchanged, while Z is increased by unity. In "gamma decay," neither A nor Z is changed—a hint that the nucleus is not an elementary particle.

In the year of my birth, 1932, the neutron was discovered. Its discovery marks another return to stage A of our cyclical history. Neutrons and protons (hydrogen nuclei) were the evident constituents of all atomic nuclei. The science of *nuclear physics* was born, whose gifts to the world were atomic weapons (the bad news) and nuclear power (the good news). The complacency represented by Gamow's remarks set in. Surely, we had at last discovered the basic constituents of matter.

Once again, a population explosion occurred. Neutrons and protons were joined by many other particles, with an equal claim to elementarity. No one in his right mind could believe that all these particles were elementary. Within the choas, patterns began to emerge. In the fifties, many competing schemes were created to organize particles into an analog of the periodic table. In 1961, a system called "the eightfold way" was invented by Murray Gell-Mann and by Yuval Ne'eman. Instead of the familiar rectangular array of the periodic table of the elements, patterns consisted of mysterious triangles and hexagons. But like the periodic table, it predicted the existence of new, and not yet observed, particles. *The particles were found,* and they had just the properties predicted.

Not only were there far too many elementary particles for them to be elementary, and not only did they seem to organize themselves into families and patterns of the eightfold way, but they displayed characteristic properties of composite, structured systems. They were not pointlike; they had measurable sizes. Their electrical and magnetic properties were distributed within a finite and measurable volume. Distributed on what? Furthermore, each so-called elementary particle had a rich structure of excitations. It seemed as if they were made of simpler ingredients which could rotate or vibrate with respect to one another.

In 1967, experiments done at Stanford University seemed to indicate that the proton consisted of several parts, or constituents. These were facetiously called "partons"—the constituents of which protons were made, if protons were made of constituents. Soon it became clear that partons were the very quarks that Gell-Mann had conjectured several years earlier.

One of the greatest successes of atomic physics of the twenties was in its ability to explain the periodic table. Properties of chemical elements could be explained, indeed computed, in terms of the quantum-mechanical theory of the nuclear atom. In a similar fashion, a theory of quark interactions emerged in the seventies. The mysterious successes of the eightfold way could be explained by the new theory of "quantum chromodynamics." It showed, for example, that three quarks would bind together to form a proton or a neutron. Strangely, the theory also says that it is impossible to produce a single, isolated quark. Just as there is no such thing as a string with only one end, or a magnet with only a north pole, there is no such thing as an isolated quark. Thus, quarks are constituents of an entirely novel kind, which may only exist as parts of a whole. Nonetheless, evidence for their existence is compelling, and the unmistakeable sign of the quark is revealed in any number of high-energy physics experiments.

Here we are again at stage A. Quarks are constituents of nuclear particles, and they show every indication of being truly elementary. Together with another class of particles, the leptons, they are today's version of the ultimate constituents of matter. Particles called leptons do not partake in the strong nuclear interactions. Like electrons, they are extranuclear particles. They are quantum chromodynamically inert.

Yet again, there has been a minipopulation explosion. The sixth quark species was discovered at CERN in 1984.* It is called the top quark. Perhaps it is the last of the quarks to be found, perhaps not. There are also six known lepton species. Together, these 12 particles are gathered together in the contemporary periodic table of quarks and leptons, or matter particles, or fundamental fermions. Are 12 basic particles too many? Does the regularity of the table indicate a deeper layer of structure?

Theoretical physicists have invented many names for the entirely hypothetical constituents of quarks and leptons: rishons, stratons, preons,

* *Author's note:* The "discovery" of the top quark in 1984 was a false alarm. It has not yet been found, but it will be.

maons, dyons, etc.—surely names to conjure with. Other theoretical physicists have conjectured that the true periodic table contains many more columns, corresponding to particles not yet discovered: squarks and sleptons, for example. And yet, there is at present no evidence at all for the existence of another layer of the onion. Quarks and leptons, as far as can be determined, are truly pointlike particles. They have not been observed to have excitations, as do atoms or nuclei, or nucleons. No evidence for squarks or sleptons has appeared, despite the hardest efforts of experimenters. Perhaps we really have come to the bottom of the barrel. Perhaps not.

In passing through the stages of our quest, from atoms, to nuclei, to particles, to quarks, the experiments have become more difficult and more expensive. Existing particle accelerators cost hundreds of millions of dollars. High-energy physics experiments are now done by groups of scientists, including a hundred or more Ph.D.'s. Let me try to explain why this is so. The key concept is *energy* which, according to quantum mechanics, is intimately related to *size*. A high-energy accelerator plays a role much like a powerful microscope. Individual particles are accelerated to high energy in order to act as probes of the microworld. The higher the energy, the finer the probe.

Normal people measure energy in human-sized units, like calories or kilowatt-hours. Particle physicists, even so-called "high-energy" physicists, deal with much smaller units, appropriate to individual particles. Our unit of energy is the electron volt (abbreviated eV). One kilowatt-hour, for example, is the equivalent of 2×10^{25} electron volts!

Each molecule of air is in rapid and chaotic motion. The mean velocity of an air molecule is approximately equal to the velocity of sound, about 300 meters per second. At room temperature, its average energy is 1/40 eV. Chemical reactions produce considerably more energy per particle. When a carbon atom burns, it releases several electron volts of energy. This is why a fire is a relatively dramatic event. Each particle of visible light, or photon, carries an energy of several electron volts. The eV is characteristic of chemistry, of optics, and of the external portion of the atom. The one electron volt accelerator is a familiar household device: the flashlight battery. It is not very expensive. Yet, it is the study of phenomena at a few electron volts that revealed the atom as a structured system. Quantum mechanics and atomic physics emerged from the observation that a flame turns yellow when the pot boils over. Such an experiment, done a bit more carefully, marked the beginning atomic spectroscopy and led ineluctably to a theory of chemistry.

The inner part of the atom is far smaller than the atom itself. Its study requires probes of much greater energy. The visible photon is useless for such fine analysis. Energies of thousands of electron volts, kiloelectron volts, or keV, are required. Such photons are known as X rays. They, too, are relatively inexpensive. Even your neighborhood dentist can afford an X-ray machine. It was by the study of X rays that Moseley, just before the First World War, discovered the secrets of the inner atom. Moseley demonstrated that chemical identity is determined by nuclear charge. He established the key concept of atomic number.

As energetic as X rays are, they are not sufficient to explore the inner structure of the nucleus itself. Particles with energies a thousand times greater are required. Nuclear properties are measured in millions of electron volts, or MeV. Once again, nature makes particles with MeV energies freely available. Such particles are produced by processes of natural radioactive decay. They were employed by Ernest Rutherford to discover the nucleus itself in 1911. It was by the use of radioactive probes that the neutron was discovered in 1932. Physics at MeV energies is still performed today. The atomic nucleus has not yet revealed all of its secrets. Small and specialized accelerators producing beams of particles with MeV energies operate throughout the world, wherever nuclear physics is pursued. A wealthy university or a developing country can (and does) easily afford to build such facilities.

To study the structure of protons and neutrons, a thousandfold giant step in energy is required. Particle physics energies are measured in billions of electron volts (or GeV). The antiproton was discovered soon after the Berkeley Bevatron was built. It created approximately 3 GeV of energy per collision.

Our contemporary synthesis, the standard model of quantum chromodynamics and the electroweak theory—with "elementary" quarks and leptons—arose out of discoveries made with experiments at several GeV of energy. Now, at last, we have come to Big Science. In the United States, four laboratories operate at the multi-GeV frontier: Brookhaven National Laboratory in New York, Fermilab in Illinois, Stanford Accelerator complex in California, and a more modest, but very productive facility, at Cornell University.

Our quest may come grinding to a halt unless we take one more giant step. To learn why quarks and leptons behave as they do and whether they are the ultimate particles of nature, we need energies of many trillions of electron volts (or TeV). Our European friends were the first to attain collision energies of half a TeV, which led them to the discovery of the W particle and the Z particle. Since 1988, the U.S. Fermilab Collider has been operating at an energy of almost 2 TeV. However, it has become evident that even this energy is insufficient to answer our most vexing questions. To this end, American high-energy physicists have set as their highest priority the construction of a giant accelerator producing collision energies of 40 TeV. This dream machine may have a circumference of a hundred miles, and will cost billions of dollars to build. Frontier physics is clearly not as cheap as it once was. Will the U.S. government build our Superconducting Super Collider, as this giant machine is known? Is it a reasonable way for U.S. taxpayers to spend their hard-earned money? Will the construction, deployment, and implementation of such an expensive facility lead to new technological and scientific advances that will contribute to the progress of modern civilization?

THE WORK OF A THEORIST: ELEMENTARY PARTICLES

What Is an Elementary Particle?

"Do we do fundamental physics to explain the world about us?" is a question that is often asked. The answer is NO! The world about us was explained 50 years ago or so. Since then, we have understood why the sky is blue and why copper is red. That's elementary quantum mechanics. It's too late to explain how the work-a-day world works. It's been done. The leftovers are things like neutrinos, muons, and *K*-mesons—things that have been known for half a century, still have no practical application, and probably never will: little mysteries like how the universe began and how it will end.

So it is that we are not trying to invent a new kind of toothpaste. What we are trying to do is to understand the birth, evolution, and fate of our universe. We are trying to know why things must be exactly the way they are. We are trying to expose the ultimate simplicity of nature. For it is in the nature of elementary-particle physicists (and some others) to have a faith in simplicity, to believe against all reason that the fundamental laws of physics, of nature or of reality are in fact quite simple and comprehensible. So far, this faith has been extraordinarily productive: those who have it, often succeed; those without it, always fail.

The field is, or can be, divided into three parts. One is particle physics and is the subject of this essay. Another is the study of gravity from Newton to Einstein and beyond and, in particular, the general theory of relativity, which every physics graduate student must learn. Lastly, cosmology is the study of the origin of the universe in the Big Bang. Particles, gravity, and cosmology seem to be quite different kettles of fish. Gravity explains the motions of heavenly bodies; cosmology deals with the birth and death of the universe; and particle physics is concerned with the most microscopic and impractical particles in the world. These three sciences are finally converging. We are heading breathlessly toward the (we hope) one and true theory, the ultimate theory of nature which will explain everything. We're not quite there yet.

In Figure 1, you see a snake swallowing its own tail—an illustration which, I'm told, has great mythological significance. It is the snake of relative sizes. The numbers here inside the snake correspond to sizes of objects from very small objects (here 10^{-18} centimeters) to very large objects (here 10^{+18} centimeters). We are at the middle of the snake, within its belly. That is because we are of thoroughly intermediate size: not very big and not very small. Running up the snake are things like mountains, the earth itself, then the entire Solar System, about a light hour in extent. Still larger things are the distance to the nearest star, the size of our galaxy, the distance to our sister galaxy Andromeda, and finally the size of the entire universe. Going down the staircase, we come

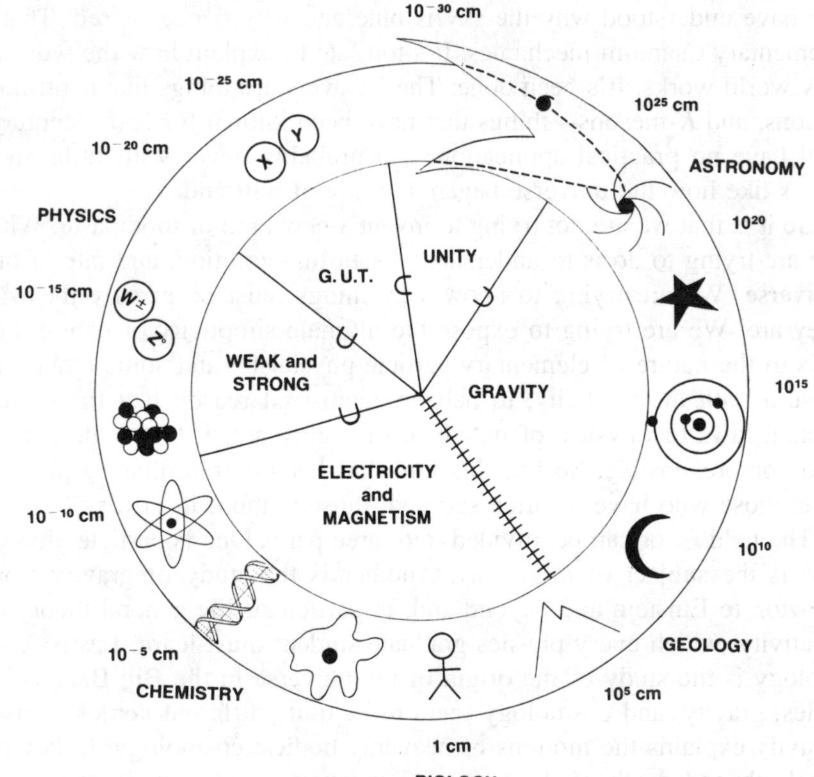

FIGURE 1. *The Great Snake of Relative Sizes.*

to fleas and lice, then to bacteria, then to strands of the large DNA molecule, then the atom itself. The atom's nucleus is smaller still, by a factor of 10,000. Smaller still are the quarks that make up individual neutrons and protons. A hundred times smaller are the W and Z particles, and smaller yet are the still-hypothetical ingredients of our grand unified theory.

Strangely enough, it turns out that when you study the physics of very small objects, that magical force gravity becomes relevant again. It's gravity which is the force that controls all these big objects. Gravity is important for the earth, our solar system, stars, and for the Galaxy. Again, gravity is important for exceedingly small objects. So the snake does, in fact, swallow its tail. The mystery, the beauty of the modern synthesis between particle physics and cosmology is the study of this particular fact: that gravity is essential to the physics of the very small and the very large.

I try to teach this connection in my course at Harvard, which is basically physics to a hostile audience. All Harvard is divided into three parts, as you may know. One part of Harvard is devoted to science, and its students are the *sciencites.* At the opposite pole are the *humanoids,* who study English, literature, and the classics. In between are the undecided: sociologists, economists, and followers of political "science": in short, the *sociopaths.* My class consists of an equal mixture of humanoids and sociopaths. I tell them the tale of the second snake, which spirals inwards, encountering ever more plausible candidates for the ultimate constituents of matter. That is the story of this essay. It is a periodic progression. Let us begin with atoms.

Once upon a time, atoms were thought to be fundamental and immutable. But there were so many different kinds of atoms, or elements, for them to all be elementary. Atoms were then shown to have periodic and regular properties as revealed in the periodic table. The success of the periodic table suggested to its inventor, Mendeleev, that the atom was a structured system made of simpler things. Half a century later, it became clear that it was. The atom is made of electrons going around a central nucleus, and this led to the quantum-mechanical view of the nuclear atom and to a new level of structure. The atom turned out to be not elementary at all. It is a complex object made of atomic nuclei and electrons.

Were the atomic nuclei themselves elementary? Of course not. Isotopes had been discovered. There are more atomic nuclei than there are different kinds of atoms, and shortly the discovery of regularities was to appear. In this case, it was the oddly integral values for atomic numbers and atomic weights that led to a new level of structure. With the discovery of the neutron in 1932, it became clear that the nucleus was made up of neutrons and protons. Earlier, it was erroneously believed that nuclei were made up of protons and electrons, the only elementary particles then known. Mesons were invented to provide the forces that hold protons and neutrons together. We were led to a new discipline—the meson theory of nuclear forces—and to a new category of elementary particles: neutrons, protons, mesons and their many cousins.

With the deployment of large accelerators in the 1960s, elementary particles entered a period of population explosion. It turned out there were hundreds of them. Just as there are hundreds of different atoms, and there are 399 different kinds of atomic nuclei that live for at least a year, so there are at least a couple hundred different kinds of elementary particles—too many for all of them to be regarded as elementary. For something to be an elementary particle, it's good not to have too many kinds of them.

Once again, order was restored in new kinds of periodic tables, in a system known as the eightfold way. Ultimately, this system led to a new vision, a new structure, which is now called the standard theory. Today's dogma is called "quantum chromodynamics and the electroweak theory." These theories appear to do an awfully good job of explaining the world about us in terms of a new level of fundamental particles called quarks.

But things have progressed to the point that there is a population explosion for them too. There are now known to be six kinds of quarks, a number which is neither threateningly large nor satisfyingly small.

The study of physics itself is a periodic phenomenon. We've gone from atoms, to nuclei, to particles, to quarks and, in each case as soon as a candidate for the elementary particle appears, it is followed by a degree of chaos (the population explosion), followed by a layer of order (the periodic properties of these potential elementary particles). Then hints of structure appear and a new candidate is offered as an elementary particle. Today, that candidate is the quark. This logical structure can also be shown as a snake—a snake spiralling inwards, seeking the center of the apple of knowledge (Figure 2).

Let's go back in time—to 1869 and the invention of the periodic table—in order to follow the structure further. In Figure 3, we see a portion of the periodic table, a fragment of some historical importance. The figure shows the dates of the discovery of some of the entries:

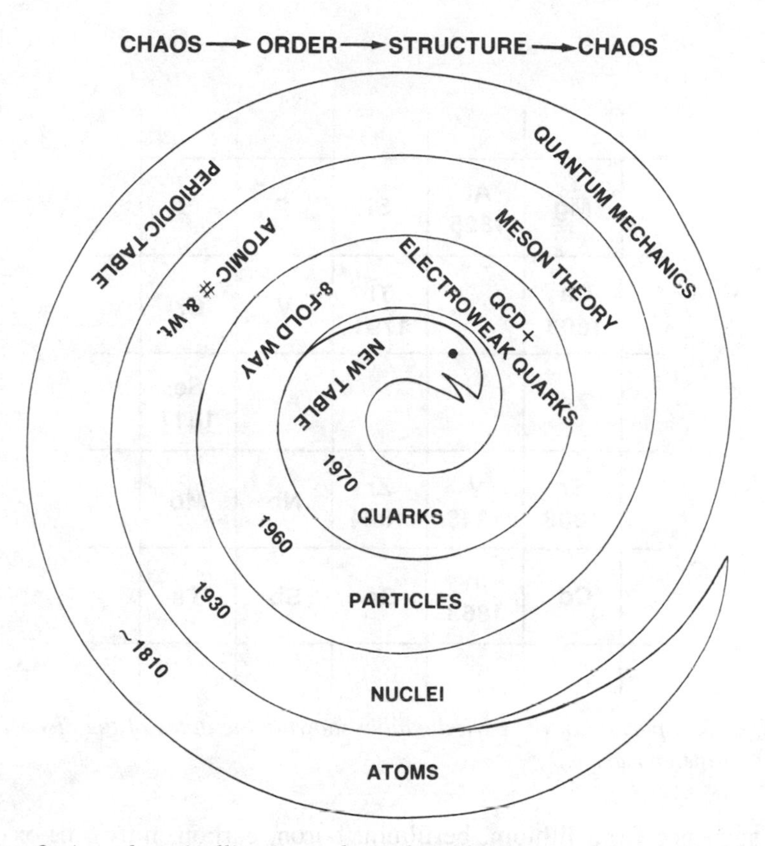

FIGURE 2. *A snake spiralling inwards, seeking the center of the apple of knowledge.*

Yttrium (Y), for example, was discovered in 1843, and Zirconium (Zr) in 1824, and aluminum (Al) was isolated only in 1825. Mendeleev, who produced his periodic table in 1869, noticed that it had a very curious feature; namely, that it had these holes, corresponding to elements that had not yet been discovered. He predicted the existence of three new elements in order to complete his table, and he predicted the properties of the missing elements. Sure enough, scandium was discovered in 1879, gallium in 1875, and germanium in 1886.

The periodic table was a mystery: How did it work? Why did it work? We didn't understand this until the 1920s or so, when it was finally explained, understood, and recapitulated in terms of fundamental atomic theory. Those who have studied chemistry know that we now understand the structure of the atom and the periodic behavior of elements—how

Mg	Al 1825	Si	P	S
Ca 1808		Ti 1791	V	Cr
Zn			As	Se 1817
Sr 1808	Y 1843	Zr 1824	Nb	Mo
Cd	In 1863	Sn	Sb	Te

FIGURE 3. *A portion of the periodic table showing the dates of the discovery of some of the entries.*

the sequence from lithium, berillium, boron, carbon, nitrogen, oxygen, fluorine, etc. moves with increasing valences and regularities of chemical combinations. We now know how these facts are explained in terms of the constituency of electrons within the atom and how that constituency itself is determined by the electric charge of the nucleus. The periodic table has been explained in terms of the heavy small nucleus (which is taken as a given by chemists), its integral atomic charge (to be precisely neutralized by orbiting electrons). The whole ensemble is subject to the laws of quantum mechanics, which date to the year 1926. Since then, the table is no longer a mystery, but useful to our chemist friends who invent charming new products like silly putty. Chemistry is no longer as much a forefront, fundamental science as it was a century ago.

Now let's turn to the nucleus, which once upon a time was the subject of a branch of fundamental physics. In nuclear physics, again we ask: Is there some kind of analog to the periodic table? What does it mean that the electric charge of the nucleus is an integer? Why is it that every atom has a mass that is an approximately integral multiple of the hydrogen atom? Indeed, with the discovery of isotopes, it became clear that all of light nuclei had almost integral atomic weights. With hydrogen defined to be one, helium weighs four, boron eleven, and carbon twelve. Poor chlorine has an atomic weight of 35.5, as far from an integer as it could be. But that was explained because it turns out that chlorine was discovered in the 1920s to consist of 75 percent chlorine-35 and 25 percent chlorine-37, both isotopes with nearly integral atomic weights.

So what do those integer weights and integer charges mean? Since 1932, we know that it means that the nucleus is not an elementary object. It's a little bean bag made up of two kinds of beans: protons and neutrons—both with approximately the same mass. The proton is electrically charged and the neutron is not. So we can close the book on the nucleus. Let nuclear physicists play their childish and dangerous games: we are concerned with the truly fundamental.

Let us go on to the study of the so-called elementary particles. When I was a lad in high school, the list of these particles was very short: it included only protons, neutrons, electrons, and the conjectured but yet unseen neutrinos. At that time, George Gamow wrote *One, Two, Three, Infinity,* one of the books that got me interested in particle physics. He wrote that there were only these four kinds of elementary particles, each with the simplicity of a mathematical point. Although four is not one, it is the best we could do—after all, four was good enough for the ancient Greeks. We had reached the bottom, Gamow thought, of our search for the basic building blocks of matter. Of course, Gamow was dead wrong, just as earlier thinkers were wrong who argued that atoms were elementary, immutable, and not made of simpler stuff. In the very year that Gamow wrote his book, the population explosion of elementary particles began with the British discovery of half-a-dozen "strange particles."

It turned out that there are not just four "elementary" particles, but literally hundreds of them. Each had an equal claim to elementarity. Murray Gell-Mann at the California Institute of Technology, and independently, Yuval Ne'eman in Israel, invented a theory that was known as "the eightfold way." It restored some degree of order to the chaotic jumble of newly-discovered particles. Their theory predicted that elementary particles should occur in well-defined multiplets consisting of eight species (forming a hexagon when suitably displayed), or of ten species (forming a triangle). Three of these arrays are depicted in Figure 4. All of the particles in the two hexagonal arrays were already known

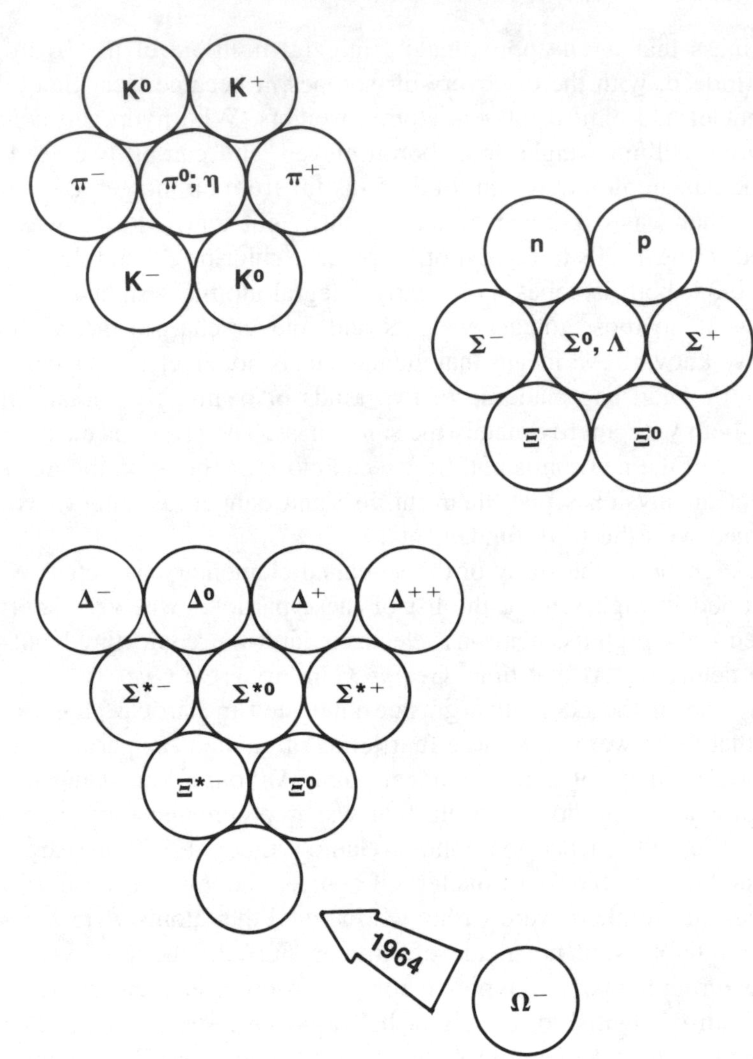

FIGURE 4. *Three of the particle arrays predicted by "the eightfold way." The misssing Ω^- particle was discovered at Brookhaven National Laboratory in 1964.*

to exist, but three of those within the triangle had not yet been seen. In 1962, the Ξ^{*0} and Ξ^{*-} were discovered, and Gell-Mann immediately predicted what the properties of the one missing particle must be.

Lo and behold! The Ω^- was sighted at Brookhaven National Laboratory by Nick Samios and his friends. It behaved just the way Gell-Mann said that it must. History repeated itself: Gell-Mann's theory was akin to Mendeleev's periodic table. At first, no one took either of these great advances seriously. Then, when the holes in the patterns were filled up, the heretics converted and the atheists found God. Gell-Mann and Ne'eman's analog to the periodic table was correct. Its predicted particles exist in nature. People began to suspect that there must be some further layer of structure. And, indeed, there is.

Today we know that these particles are not as elementary as they once seemed. They are now known to be made up of quarks and satisfy the quark commandments. Observable nuclear particles are of these and only these varieties: Baryons, each of which is made of three quarks; antibaryons, each made of three antiquarks; and mesons, each containing one quark and one antiquark. Gell-Mann's mystic triangle consists of particles which are made of subparticles. One particle, for example, (Δ^{++}) is made of three up quarks; another (Δ^{-}) composed of three down quarks; (Σ^{*0}) in the middle is made up of one up quark, one down quark and one strange quark; and the strangest of all particles (Ω^{-}) is made of three strange quarks. (I must apologize for the names of the quarks, which were given to them by Gell-Mann.) Almost all of the matter in this room consists of quarks, with a tiny residue made up of electrons. Nucleons and mesons, once thought to be ultimate forms of matter, are now recognized to be lowly composites. ''Elementary particles'' are not at all elementary, and their study has been reduced to a kind of chemistry: mere child's play.

The proton is made of two ups and a down. The neutron is made of two downs and an up. It follows that quarks bear fractional electric charges. The up quark carries a charge of ⅔ and the down quark −⅓. If an isolated quark is to be observed, its characteristic and striking feature is its fractional electric charge. Many experimenters have searched for the fractionally charged quark. Aside from a group in California that persists in seeing what it must not see, the quark has not been seen in the laboratory. It has not and it cannot. Today's theory of quarks demands that quarks cannot be produced in isolation. They only exist as parts of a whole and they cannot exist as particles in their own right.

For example, imagine attempting to remove a quark from a hadron, using, I suppose, a quarkscrew. The force holding the quarks together does not diminish with distance: to remove the quark, the spring must be stretched, and at a certain point the spring will break. When this happens, a new quark-antiquark pair is produced. In this procedure, we fail to isolate a quark. Rather, we produce a meson.

At last we have come to today's candidates for the ultimate building blocks of matter: the latest ''periodic table,'' that of quarks and leptons. Four different particles are required to build and to operate Spaceship

Earth, and its gigantic power plant, the sun. Two kinds of quarks, up and down, are the ingredients of neutrons and protons. Together, they represent the vast majority of the mass of sun and earth. Electrons, combined with these aggregates of neutrons and protons, called atomic nuclei, complete the construction of the atom. Finally, there is the elusive neutrino. While it is *not* an atomic constituent, it does play a vital role in the process known as beta decay. In the early universe, beta decay made possible the synthesis of heavier nuclei from primordial hydrogen. And it is this mechanism that enables the sun to operate its nuclear furnace and to warm the earth and its inhabitants. Without neutrinos, Spaceship Earth, if it would exist at all, would be a frozen ball of solid hydrogen, and life could never have evolved. Just four fundamental particles were enough to fashion all the wonders of our earthly paradise: *two quarks* with electric charges ⅔ and − ⅓, one negatively charged *electron,* and its accompanying *neutrino.*

Four is a small number, but unfortunately it is not as simple as all that. Particle physicists have made a great leap sidewise. They have discovered particles that are in no sense constituents of ordinary matter. And, these curious particles came in families that are tantalizingly similar in structure to the first family of four "relevant" particles.

The great leap sidewise began in 1938 with the sighting of a new particle in cosmic rays. Much like the electron except that it is about 200 times heavier, this particle is now called the *muon.* "Who ordered that?" said Isadore Rabi upon hearing about this discovery while lunching in a Chinese restaurant, little realizing that the muon was the first member of what was to be a second family of fundamental quarks and leptons.

When unstable particles decay, they often produce an electron or a muon accompanied by a neutrino. In 1961, experimenters at Brookhaven National Laboratory proved that the neutrino accompanying a muon was a different particle than that accompanying an electron. For the discovery of the muon-neutrino, the second member of the second family, Leon Lederman, Mel Schwartz and Jack Steinberger were awarded the 1988 Nobel Prize in physics.

Certain particles that were first seen in cosmic rays in 1947 behave peculiarly and are known as *strange particles*. Today, we know that a strange particle is nothing other than a particle containing a strange quark. The strange quark is simply a heavy cousin of the down quark. Unlike the up and down quarks, it is not an essential component of earthly matter. In 1963, when Gell-Mann invented the notion of quarks, the second family was complete except for the missing heavy relative of the up quark. Again, there was a hole in the table, a yet-to-be discovered particle that would complete the pattern. Bjorken and I conjectured the existence of the *charmed quark* in 1964, but we had to wait a decade for its discovery.

With the discovery of charmed particles in the 1970s, the table seemed to be complete. There were two symmetrically related families of fundamental fermions: the relevant and the irrelevant. Or, so we thought.

Martin Perl, working at the Stanford electron-positron collider known as SPEAR, revealed the first evidence for the existence of a *third* family of fundamental particles. He discovered the tau lepton, a heavier version of the muon. In the years since, indirect evidence compels us to believe that the tau has its own associated neutrino. Moreover, Lederman and his co-workers at Fermilab have proven that there exists a fifth quark, variously called the bottom or beauty quark, or perhaps, the bicentenary quark in honor of the year and the nation of its discovery. It is like the down or strange quarks, but it is much heavier. Once again, the table is almost complete. There are six known leptons, but only five known quarks. A heavier relative of the charmed quark must exist, but it has not yet shown up despite the tireless efforts of experimenters worldwide. It is called the top or truth quark, and hopefully not the tercentenary quark.

So much for the matter particles. Now we come to the forces to which they are subject and by which they form the observed and observable manifestations of nature. Physicists know of only four fundamental forces. Gravity is perhaps the most familiar of the forces. It keeps the planets in their orbits about the sun, and it almost succeeds in trapping mankind upon the surface of our planet. Almost, but not quite, since after all, we have managed to walk on the moon and return unscathed. In today's study of the microworld, gravity is practically irrelevant. Being the weakest of the forces, it has no discernible effect on elementary particles.

One force, and one force alone, is responsible for almost everything that we can see, feel, smell, taste and feel. It is called electromagnetism. It is the naked force of electromagnetism that holds electrons close to their nuclei. All chemical and mechanical forces result from electromagnetism. It is all we need to know until we probe the structure of the nucleus itself. The quantum theory of electromagnetism is an immensely successful discipline. It allows us to compute observable phenomena to an accuracy given by many decimal digits. It is the paradigm for a theory of all natural phenomena. Quantum electrodynamics deals with electrons and photons, or particles of light. The photon is the agent that mediates the electromagnetic force.

From the study of particles and nuclei, the existence of two other forces has been revealed. These are called, simply, the strong force and the weak force.

The theory of the strong force was developed in the 1970s and is known as quantum chromodynamics. It is the strong force that holds quarks together to form neutrons and protons. The strong force is mediated by a type of particle called the *gluon,* which can be emitted or absorbed only by quarks. Like quarks, gluons cannot exist in isolation. Their realm of existence is limited to the interior of elementary particles. Evidence for the reality of gluons was produced in Germany in 1981, where gluon jets were first observed at the large German electron-positron collider PETRA.

Today's theory of the weak interactions emerged in the 1960s. A consistent theory of these forces required that the weak force be unified with electromagnetic interactions. This once very speculative ''electroweak theory'' made three spectacular predictions. It said that there must exist in nature a new variety of phenomena called ''neutral currents.'' Sure enough, these predicted effects were discovered to exist. They were first observed at CERN in 1973, and soon thereafter, they were seen by Rubbia's group which was then at work at Fermilab.

In addition, the electroweak theory demanded the existence of new particles to mediate weak interactions, particles which are known as W and Z. These particles are so heavy that their existence could not be established at any existing accelerator. That is why my colleague Carlo Rubbia proposed the construction of a new facility dedicated to the discovery of these tantalizing particles. For whatever reasons, the U.S. was unable, or unwilling, to construct the facility in a timely fashion. But our European colleagues immediately set about to build Carlo's dream machine—the CERN Proton-Antiproton Collider. Soon after the machine was built, Carlo was able to establish the existence of both the W and Z. For this great discovery, Rubbia shared the 1984 Nobel Prize in physics. The other half went to the Dutch engineer, Simon Van Der Meer, whose genius made possible the successful deployment of the CERN collider.

What are our elementary particles today? We have twelve matter particles (quarks and leptons) and four force particles (photons, gluons, W's and Z's). Aside from the top quark, there is still one exotic particle missing and remaining to be discovered—a curious object known as the Higgs boson. All told, our bestiary contains a total of 17 "elementary" particles. Not only does our theory involve 17 particles, but it contains 17 arbitrary and adjustable numbers or parameters. Should the ultimate theory involve so many different kinds of particles? Should our creator be forced to adjust 17 dials in order to "tune us in" properly? Hell no! Surely, we do not yet have the last word. Surely, Nature's bag of tricks is not yet exhausted, and she will amaze us once again. Surely, there is a lot more to be done in the realm of fundamental physics. No scientist with good taste can accept 17 truly fundamental particles. The snake is not yet at the core of the apple.

Let us return to the history of our discipline for guidance. The ancient Greeks believed in just four fundamental elements: fire, water, earth, and air. The number was small enough but the theory, such as it was, was all wet. Lavoisier, when he developed modern chemistry two centuries ago, was able to identify about 30 distinct chemical elements. In the intervening years, the list has expanded to 110. Indeed, there are so many different chemical elements that chemists have stopped bothering to give them names. Elements are often referred to by their atomic numbers, just as taxpayers are known by their Social Security numbers.

It is interesting to study a plot of the number of known chemical elements over the past two centuries. The number increases smoothly from 30 to 110, but not quite linearly. Periods of explosive growth in the number of elements reflect the development of new technologies: electrochemistry, spectroscopy, and radiochemistry. Nonetheless, I was taught that there are 92 chemical elements. But only for a brief moment were the known elements exactly 92. Somehow, it had become accepted dogma within the closed circle of high-school chemistry teachers. It is rather like the affectation of physicians to pronounce *centimeter* in a pseudo-French fashion, "sontimeter."

Compare the growth in the number of known elements to the growth in the number of identified "elementary" particles. In 1950, there were half a dozen, while today there are more than 100. Again, the growth curve shows several spurts corresponding to the development of bubble chambers in the early 1960s, and of electron-positron colliders in the mid-1970s. Population explosions of atomic elements and of elementary particles were indications that they could not be taken seriously as truly elementary particles. Note, however, how rapidly things happen today. The growth of the number of elements took two centuries, while more recent population explosion took a mere two decades.

The growth in the number of quarks has been more modest. Originally, when quarks were invented, there were only three quark flavors. Today, there are six. Is this enough of a population explosion to convince us that quarks are *not* elementary? I think that it is too early to tell. What of our list of 17 elementary particles? While I was a graduate student, only some 41 percent of these had been found. Today, 15 of them have been shown to exist; we seem to be 88 percent done. Surely, however, our list is subject to change. It has always been true that today's elementary particle turns out to be tomorrow's composite system.

Today, we have a plausible list of elementary particles, and a powerful and predictive theory: quantum chromodynamics and the electroweak theory. Only recently were the predicted W-particle, Z-particle, and top quark found. One might expect us to be proud and jubilant with our successes. Quite the contrary. We pray that our experimental colleagues will prove the theory wrong. We live in anticipation of a surprising experimental discovery, something that does not fit in with our theory. We need a loose end in order to fashion the next development and a new and better theory. Our progress depends upon being proven wrong. Carlo Rubbia has made a series of outstanding discoveries, discoveries which confirm our theory. With baited breath, we await the unexpected, proof that our vision of nature is not yet complete.

How do we know whether quarks and leptons are truly elementary? What is the essential property of a composite, as opposed to an elementary, system? One answer is the existence of quantum excitations, or energy levels. It is simpler than it sounds. In a composite system consisting of discrete parts, one substructure may move with respect to the others. It may rotate, or vibrate, or otherwise wiggle and jiggle inside. These motions are what give rise to a spectrum of energy levels. Atoms and molecules have them. Indeed, the study of atomic energy levels (or spectroscopy) was an essential precurser to our understanding of the atom as a structured system. The atomic nucleus, similarly, possesses a number of excited states, or energy levels, in which its component nucleons may enjoy complex motions. Even the nucleon itself has a complex and well-studied array of energy levels, a consequence of the fact that it is a complex and structured system containing three quarks. So it seems as if the vital question is whether or not quarks and leptons are subject to internal motions. So far, evidence is negative. Neither quarks, electrons, nor neutrinos have shown any evidence at all of inner structure. Perhaps they are truly elementary after all. Or perhaps we must study them more carefully with better instruments and at higher energies.

As we strike more deeply into the inner structures of matter, we must use probes of higher and higher energy. This is a consequence of quantum mechanics, and of the conjugate relationship between distances and momenta. Light, for example, has a characteristic size, or wavelength, which far exceeds the size of an atom. We cannot study the inner structure of atoms with visible light. We need shorter wavelengths, and higher energies. The inside of the atom was revealed by means of X rays—light-like particles thousands of times smaller and more energetic than visible photons. These particles each have an energy of *thousands* of electron volts.

The study of the nucleus requires even larger energies. Early nuclear physics depended upon the existence of natural radioactivity, which makes available energies of *millions* of electron volts. Rutherford used such particles to discover the nucleus in the first place. He noticed that alpha particles would sometimes be scattered backwards from a piece of gold foil. Rutherford said, years later, that it was as if a cannon ball fired at a piece of tissue paper came back at you. From this experiment he concluded that most of the mass of the atom was concentrated in a pointlike nucleus.

It is interesting to note that an experiment much like Rutherford's led to the first compelling evidence for the existence of quarks. It took place in Stanford in the late 1960s at what was then the largest accelerator in the world. Very high-energy electrons were made to collide with hydrogen atoms. Again, too many of the electrons were observed to be deflected by large angles. Clearly, there were small and hard scattering centers within the proton itself. At first, these parts of the proton were called "partons." Today, we know that the parton is a quark.

It was through the study of natural radioactivity that the neutron was discovered in 1932. With the neutron and proton as fundamental building blocks, a rational doctrine of nuclear physics emerged. Nuclei had almost integral atomic weights because they were made up of so many nucleons. These were held together by a new kind of force, quite distinct from gravity or electromagnetism. Yukawa, in the 1930s, speculated that this force was mediated by a hypothetical particle intermediate in mass between the nucleon and the electron.

Many of the early discoveries in particle physics were accomplished by the study of cosmic rays. These mysterious particles are exceedingly energetic, and they arise from distant parts of our own galaxy. Cosmic-ray experiments yielded the first evidence for the existence of antimatter: the positron, antiparticle to the electron, was observed in 1932. In 1938, a particle was observed in cosmic radiation which was intermediate in mass. Was it Yukawa's meson, the glue that holds nucleons together? As it turned out, it was not. It was the mysterious heavy electron we call the *muon*. Yukawa's particle was observed by a cosmic-ray physicist in 1947, along with a number of curious and unanticipated particles known as strange particles.

While cosmic rays have an enormous amount of energy, the trouble is that there really aren't very many of them. From the point of view of human survival upon earth, this is fortunate. However, it means that particle physicists needed an artificial source of high-energy particles. More and more, the study of high-energy physics depended upon the construction of larger and larger accelerators.

As accelerators were deployed, the pace of discovery quickened. The first excited state of the nucleon was discovered in 1954, the antiproton was first observed in 1955, and the existence of at least two kinds of neutrinos was demonstrated in 1961. The Ω^- particle, whose existence proved the validity of the eightfold way, was detected at Brookhaven National Laboratory in 1964. Indeed, practically all we know about elementary-particle physics emerged from experiments done at large accelerators. An exception, which perhaps proves the rule, is the observation of parity violation in the mid-1950s by Professor C. S. Wu and her collaborators at Columbia. It may have been the last of the great "tabletop" experiments in elementary-particle physics.

As larger and larger accelerators were constructed and put to work, we could probe ever more deeply into the subnuclear mystery. With the machines of the 1970s, both the fourth (charmed) and fifth (beauty) quarks were discovered, as well as the third charged lepton, the heavy cousin of the muon, known as the tau lepton. Notice that all the accelerator-related discoveries we have mentioned took place in the United States. Particle accelerators were a U.S. invention, and U.S. scientists employed them most effectively. All that has changed. In the last decade, our West European friends have more than caught up.

In the early 1970s, there was a race for the discovery of neutral currents, those predicted effects that would confirm the validity of the electroweak theory. It was a race between CERN (Europe's major particle physics laboratory) and Fermilab. Europe won the race by a nose—it was their first really great triumph.

Carlo Rubbia was one of the first to sense the significance of the event, for he was one of the leaders of the losing team. He knew that the great test of the theory was the search for the W and Z bosons, a search which could not be accomplished with any existing accelerator. With two American colleagues, he proposed the construction of a proton-antiproton collider, a daring modification which could be implemented at the existing accelerators at CERN or at Fermilab. CERN responded to the challenge with alacrity and dedication. The CERN collider began operating in 1981 while the larger but slower Fermilab collider did not start up until 1987. That is why the W boson was discovered in 1982 and the Z boson in 1983, both in Europe.

Our collider is now in operation, but it is far too small to do the trick. U.S. high-energy physicists do not intend to give up the competition. Their number-one priority is the construction of an accelerator far more ambitious than any on the drawing boards. We dream of the Superconducting Super Collider (the SSC), a colliding-beam facility with about 40 times higher collision energy than is now available at Fermilab. This machine, now being built in Texas, will be the largest scientific facility ever—it will be 52 miles in circumference. What do we imagine we shall learn with this great machine?

In the past decade, we have come to a plateau in elementary-particle physics. We have developed an apparently complete list of 17 "elementary particles." Our theory of their interactions is apparently complete and correct. No experiment is in contradiction with the standard theory of electroweak and color force.

Yet, there *are* new things to discover, if we have the courage and dedication (and money!) to press onwards. Our dream is nothing else than the *disproof* of the standard model and its replacement by a new and better theory. We continue, as we have always done, to search for a deeper understanding of nature's mystery: to learn what matter is, how it behaves at the most fundamental level, and how the laws we discover can explain the birth of the universe in the primordial Big Bang.

The Hunting of the Quark

People react to nature's glorious and bewildering display of wonders somewhat as they do to a magic show. Some sit back and enjoy the performance while others are compelled to search for rational explanations. Scientists are those who spend much of their time asking how nature's tricks work, and among these are an elite and sometimes snobbish few who pursue an ultimate question: What are the most basic constituents of matter, and how are they put together? Today, such scientists are called elementary-particle or high-energy physicists; at other times, they were known simply as physicists, or as natural philosophers, or even as alchemists.

Elementary-particle physicists follow the traditional patterns of scientific inquiry—they make reasoned conjectures about the realm of reality they hope to understand, and then attempt to verify their conjectures with experimental observations. The conjectures require great imagination and the observations occur in highly sophisticated fashion, for none of the particles at the most basic levels of matter can be seen in any ordinary sense. They are simply too small. Some of them may be viewed indirectly as they pass through various sensitive devices—they leave visible tracks in bubble chambers, produce flashes of light in scintillation chambers or make sparks in spark chambers. Others are detected only when we see them "decay"—disintegrate spontaneously to reach a more stable condition.

The field of elementary-particle physics has undergone revolutionary upheavals several times in the last century, and it is possible that we are on the verge of another breakthrough. Theoretical ideas and experimental discoveries have recently begun to come together in impressive ways. We appear close to developing a unified theory to describe the ultimate structure of matter. Current events are associated with an evocative growth of terminology: Words like ''charm'' and ''strangeness,'' phrases like ''the eightfold way'' and ''quark confinement by colored gluons'' abound in the professional literature, and even find their way into popular media. Yet the reasons for the excitement among elementary-particle physicists may seem totally obscure to the nonscientist.

What follows is a layman's tour through the jungle of elementary-particle physics. Some very familiar denizens are described, as well as some bizarre new species that have been conjectured but not yet observed. No attempt at completeness has been made, for who can tell what manner of beast will reveal itself next?

John Dalton, a founding father of atomic theory, wrote in 1810: ''We have endeavoured to show that matter, though divisible in an *extreme* degree, is nevertheless not *infinitely divisible*. That there must be some point beyond which we cannot go in the division of matter. The existence of these ultimate particles can scarcely be doubted, though they are probably much too small ever to be exhibited . . . I have chosen the word *atom* to signify the ultimate particles in preference to *particle, molecule,* or any other diminutive term, because I conceive it much more expressive; it includes in itself the notion of *indivisible,* which the other terms do not.''

Not until the present century did it become known that the atom—the building block of molecules that form all ordinary matter—is not as fundamental as Dalton thought. Far from being elementary, the atom is a complex structure somewhat analogous to the Solar System. It consists of a small, dense nucleus surrounded by a much larger cloud of *electrons*. Most of the atom's mass is concentrated in the nucleus, which also carries a positive electric charge. Each electron carries a negative electric charge. The phenomena of electricity and magnetism we see every day arise from the behavior of the light and mobile electrons. The positive charge of the nucleus is always such that it may be (and usually is) exactly balanced by the negative charge of the surrounding electrons.

Four centuries ago, Newton explained how the force of gravity keeps the planets in orbits about the sun; he thus enabled us to predict events like eclipses with great precision. In a similar fashion, the formulation of quantum mechanics 50 years ago explained how electrical force holds the atom together; it allows us to predict the properties of atoms in considerable detail.

Not even the atomic nucleus itself, which determines the chemical nature of the atom, is a fundamental, indivisible constituent of matter. The discovery of radioactivity—ejection of particles from a disintegrating nucleus—showed that one atomic nucleus could turn into another. The alchemists' dream of turning one element into another could be realized in fact.

The tiniest nucleus of all is that of hydrogen, and it is called the *proton*. The hydrogen atom consists of one proton and one electron. In the 1920s, it was hoped that all atoms were made out of just protons and electrons. That would make it simple to describe atoms by determining the number of electrons, whose negative charges imply an equivalent number of positively charged protons, which account for an atom's weight (electrons have negligible weight). For example, an atom of nitrogen has seven electrons in its cloud. Could its nucleus simply consist of seven protons, somehow stuck together? If this were so, the nitrogen atom should weigh seven times more than the hydrogen atom—but in fact, it weighs about 14 times more. Thus, it was said, the nitrogen atom must consist of 14 protons and 14 electrons, half the electrons to be found inside the nucleus, the others remaining outside, forming the cloud. But the notion of electrons within the nucleus violated some fundamental principles of magnetism, as well as other rules of physics, and no sensible theory could be found.

The situation was saved in 1932 with the discovery of the *neutron,* a particle with a mass close to that of the proton but with no electric charge. At first, the existence of the neutron caused some befuddlement. Could it be a particle composed of a positive proton and a negative electron bound together? Or was it more correct to say that the neutron is the fundamental particle, and the proton is best described as a neutron with an extra measure of positive charge?

It was soon agreed, however, that protons and neutrons (called, generically, *nucleons*) were the sole constituents of nuclei, and that they were *equally* fundamental. To explain how the nitrogen atom can weigh 14 times as much as the hydrogen atom, one can say that the nitrogen nucleus contains seven protons and seven neutrons, which weigh about the same as protons.

There was no longer any reason to put electrons inside the nucleus. This picture made sensible predictions possible. For a time, it seemed that there were just three elementary particles: protons, neutrons and electrons. With electrical force, to bind electrons to their nucleus, and a conjectured nuclear force, to keep nucleons packed tightly in the nucleus, it became possible to explain the structure of all known matter.

The electron continues to this day to be regarded as a truly elementary particle. It has an electric charge, which means that when it moves it can generate disturbances known as electromagnetic waves. These waves can carry energy and information from one place to another. In a TV transmitter, for example, the motion of electrons produces waves that are picked up by antennas and converted into sounds and images in our homes. Visible light and X rays are also thought of as electromagnetic waves. Electromagnetic waves can sometimes act as if they were particles. For example, light behaves like a particle when it hits an atom causing an electron to be ejected. This particle of light is called a *photon*. The electromagnetic force that holds an atom's electron to its nucleus can be thought of as the continual exchange of photons between the two bodies. We say that the electromagnetic force is "mediated" by the photons.

One of the most successful and far-reaching developments of twentieth-century physics is the theory of the interactions between electrons and photons called quantum electrodynamics. Its predictions agree with experimental data to an uncanny level of precision. For the past 25 years, physicists have lamented the lack of a theory describing the interactions among protons and neutrons in the nucleus with comparable power and precision.

P. A.M. Dirac first formulated the theory of the electron in 1928, and central to his theory was the predicted existence of a positively charged electron with the same mass as the ordinary negatively charged electron. Such a particle was discovered in 1932 and is called the *positron*. On being awarded his Nobel Prize in 1933, Dirac said: "The theory of electrons and positrons which I have just outlined is a self-consistent theory which fits the experimental facts as far as is yet known. One would like to have an equally satisfactory theory for protons. One might perhaps think that the same theory might be applied to protons. This would require the possibility of existence of negatively charged protons forming a mirror-image of the usual positively charged ones. . . ."

Twenty-two years later, the negatively charged antiproton was produced and observed at the world's then-largest particle accelerator at Berkeley. Now it is believed that there is an antiparticle with opposite electrical properties corresponding to each kind of particle.

No such antiparticles are found on earth outside physicists' laboratories, for it is an essential property of *antimatter* that it is annihilated by ordinary matter.

In 1934, the Japanese physicist Hideki Yukawa suggested that there might be a profound analogy between electric forces and nuclear forces. He argued that there should exist a particle to mediate the nuclear force just as the photon mediates the electric force. He called this hypothetical particle the mesotron, because it had to have a mass somewhere between that of the electron and the proton. The name was subsequently truncated to *meson.* Now the word meson refers to a whole class of particles and the particle that Yukawa predicted is called the *pion.*

The pion seemed to have been discovered in 1937, in the wake of cosmic rays. Cosmic rays are very energetic particles, usually protons, traveling through outer space. Occasionally they impinge on the earth's atmosphere, colliding with atoms in the air and producing new particles. Before the advent of large-particle accelerators, our only window to the world of high-energy subnuclear phenomena was the study of these collisions.

Most physicists believed that the particle observed was Yukawa's pion, and that his meson theory had been triumphantly vindicated by experiment. Alas, this was not the case. The particle that was discovered turned out to be the *muon,* apparently an obese electron, which has absolutely nothing to do with the nuclear force. Yukawa did, however, have the last laugh. The particle he predicted was finally discovered in 1947.

As physicists have deepened their insight into the nature of subatomic particles, they have found it useful to develop more sophisticated ways to classify them (see Figure 1). They are now divided into two broad classes: *hadrons* and *leptons*. Hadrons are particles which share in the nuclear force and interact powerfully with atomic nuclei. Leptons are particles which are immune to the nuclear force and can penetrate nuclei freely without interacting much with them.

Hadrons are divided into three smaller classes: *baryons, antibaryons* and mesons. Baryons are nucleons—neutrons, protons and some other nuclear particles—that satisfy the exclusion principle developed by physicist Wolfgang Pauli. This principle states that no two baryons may behave in the same way at the same time. Antibaryons are the antiparticles of baryons, and also obey the exclusion rule. Mesons, including the pion, are hadrons that are not constrained by the exclusion principle—more than one of them may behave in the same way at the same time.

There are four catalogued species of lepton: electrons, muons and two kinds of *neutrinos* that are associated with each. Neutrinos are mysterious and elusive particles not only free of nuclear forces, but of electrical forces, as well. They were conjectured to exist by Pauli in 1930, but were not found until 1957. Leptons satisfy the Pauli principle too.

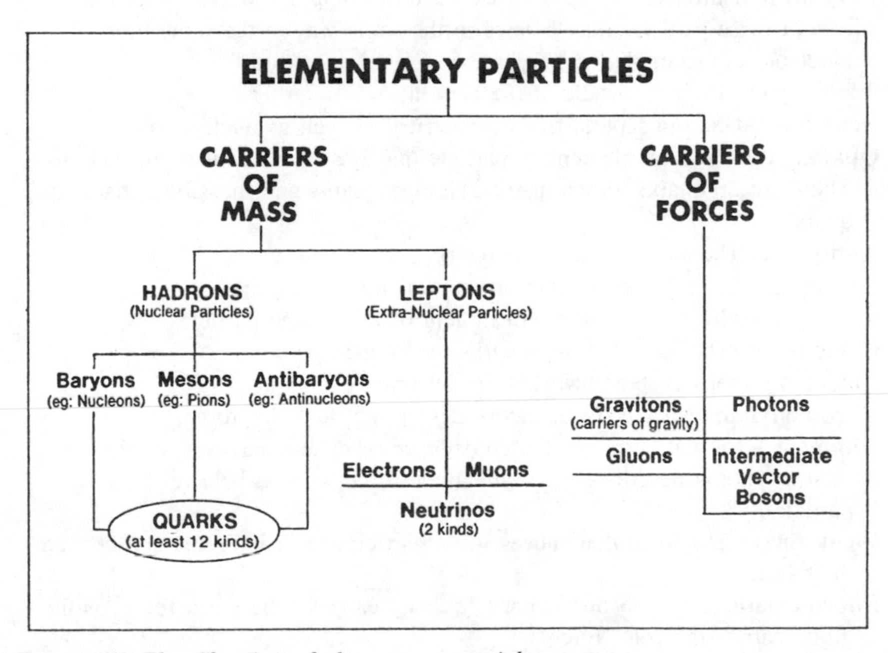

FIGURE 1. *Classification of elementary particles.*

TABLE 1. From the Microworld, in Order of Their Appearance (or Nonappearance)

Atom: A small, dense nucleus surrounded by a much larger cloud of electrons.

Electron: A truly elementary particle; it is light in mass and carries a negative electric charge.

Proton: A particle found in the nucleus; it carries a positive electric charge.

Neutron: A particle found in the nucleus; it has no electric charge.

Nucleon: A proton or a neutron.

Photon: Electromagnetic force when it behaves as a particle, rather than as a wave.

Positron: Antiparticle of the electron. A particle with the mass of an electron but with an opposite charge.

Antimatter: Matter composed of antiparticles whose electrical properties are the reverse of those of conventional particles.

Meson: One of the three classes of hadrons.

Pion: The lightest variety of meson. It is produced in the collisions of nucleons.

Muon: A particle very much like an electron but about 200 times heavier.

Hadron: A particle subject to the nuclear force.

Lepton: A particle immune to the nuclear force.

Baryon: A hadron satisfying the "exclusion principle," which states that no two of these particles may behave in the same way at the same time. Nucleons are examples of baryons.

Antibaryon: The antiparticle of the baryon.

Neutrino: A kind of lepton, free of electrical as well as nuclear force.

Quark: A conjectural elementary particle that is a constituent of all hadrons. There are up quarks, down quarks, strange quarks and, possibly, charmed quarks.

Antiquark: The antiparticle of the quark.

Strange particle: A kind of hadron unusual for its long life.

Strange quark: An obligatory constituent of all strange particles.

Color force: The force binding quarks together in particles.

Color: An unobservable but necessary attribute of quarks.

Chromodynamics: A body of theory describing the color force.

Gluon: A term that describes the color force when it behaves as a particle (just as photon describes electromagnetic force when it behaves as a particle).

Weak force: The force that causes some particles to disintegrate and change their identity.

Intermediate vector boson: A particle that "carries" the weak force, as the gluon carries the color force.

Charm: The abstract yet observable attribute borne by the fourth quark.

Charmonium: A hadron composed of one charmed quark and one charmed antiquark.

Charmed hadron: A hadron containing only one charmed quark or antiquark.

An expensive way to find out what watches are made of is to bang two of them against each other and examine what falls out. This is one of the few techniques we have to study the structure of subatomic particles, and it is for this reason that "atom smashers," or, more properly, particle accelerators, are built. In these devices, streams of protons or electrons are manipulated through electric and magnetic fields in order to increase their velocity to a point approaching the speed of light. Then they are ejected and allowed to collide with matter at rest. Sometimes these collisions result in the ejection of more particles, such as neutrons, which may then be used to bombard other targets.

When a neutron hits a proton target, the neutron may be converted into a proton and a pion. Or, sometimes the target proton becomes a neutron and a pion. Can we regard a neutron as a combination of a proton and pion, and at the same time a proton as a combination of a neutron and a pion? To make matters more confusing, sometimes the collision will result in the production of a third nucleon—an unthinkable consequence in watch-watch collisions.

These phenomena led to the notion of a nuclear democracy, wherein all of the subnuclear particles are regarded as equally fundamental. Loosely speaking, we say that each type of hadron is made up of combinations of the others. Proponents of this view do not see the need for the existence of a small set of ultimate building blocks from which all the hadrons may be built.

In the 1950s and 1960s we witnessed a population explosion of hadrons. As larger and more energetic accelerators were put into operation, more and heavier hadrons were discovered in the debris of particle-particle collisions. Today, there are about 200 known hadrons. Although most physicists would agree that all of these particles are equally elementary, they would also agree that they are probably not very elementary at all.

Patterns emerged among the hadrons. Rules governing these patterns were put forward in 1961 in a theory called "the eightfold way," which demanded that the hadrons fit into well-defined families. According to these rules, all known hadrons fit into families with either one member, eight members, or ten members.

Perhaps the hadrons are not fundamental at all but are composed of simpler things. In 1963, Murray Gell-Mann and George Zweig conjectured that there were such hadronic constituents, and named them *quarks* from an obscure line in *Finnegans Wake* (''Three quarks for Muster mark''). Baryons were assumed to be made up of three quarks; antibaryons were assumed to be made up of three *antiquarks,* and mesons were assumed to be made up of one quark and one antiquark. No other combinations of quarks corresponded to observed particles. It was possible to distinguish between two types of quarks on the basis of their different masses and electric charges, and these were called *up quarks* and *down quarks*. The two kinds of quarks were sufficient to construct neutrons and protons. Everything in the workaday world is made up exclusively of these two kinds of quarks, along with electrons.

But particle physicists know of other kinds of matter. In the 1950s, a number of hadrons were discovered in the debris of cosmic-ray collisions. They were uncommonly long-lived—they took a longer time to disintegrate than did most such hadrons, and they would do so in surprising and varied ways. They became known as *strange particles*. Today, they are routinely produced and studied at large accelerators. In order to account for these particles, it is necessary to postulate the existence of a third kind of quark, the *strange quark*.

All told, just these three kinds of quarks—up, down and strange—were enough to construct all the hundreds of hadrons that had been discovered until quite recently. Indeed, what was embarrassing about this theory was how well it worked. Its quantitative successes in describing details of hadron structure far exceeded what was expected from such a naïve picture.

There was one big problem: the quark itself had not been found. Elaborate searches were launched at many laboratories. Despite the skill and devotion of experimentalists all over the world, not one quark was seen. Not to be dissuaded from their elegant and useful construct, theoretical physicists have come up with a remarkable new theory in which the quark *in principle* is unobservable. It simply cannot be detached from the hadron of which it is a part. A meson made up of quark-antiquark is considered to be somewhat like a magnet. Any attempt to isolate the north pole of a magnet from its south pole is doomed to failure. Cut the magnet in two, and each part becomes a complete magnet with both a north and south pole. Similarly, any attempt to separate the constituents of a meson leads to the creation of a new quark and antiquark. Instead of isolating the quarks, we simply end up with two mesons.

In our search for deeper levels of the structure of matter, we have encountered molecules made up of atoms, atoms made up of nuclei and electrons, nuclei made up of hadrons, and, finally, hadrons made up of quarks. Quarks seem to be truly fundamental, for how can we learn the structure of a particle that we cannot produce? Perhaps the impossibility of finding quarks is nature's way of letting us know we have reached the end of the line.

Gravity is the only force that matters for the motions of planets and stars. For smaller things like us, gravity is important but electromagnetism plays the central role. Everything we see, hear, taste, touch and smell is but an indirect consequence of the underlying electrical structure of matter. Life itself is ultimately an electromagnetic phenomenon, albeit an exceedingly intricate one.

A third force of nature is needed to hold nucleons together in atomic nuclei. It is not a consequence of electromagnetism. Once, this nuclear force was thought to be an elementary force analogous to electromagnetism, with mesons as the most basic nuclear "glue." Today, we think that both mesons and nucleons are composite systems made of quarks. The nuclear force is regarded as a mere indirect manifestation of a more basic *color force* responsible for the permanent entrapment of quarks. *Color* is used in this new technical sense with no relation to the ordinary meaning of the word.

The theory behind the color force is complex, and not yet completely understood. Some call it *chromodynamics*. An essential requirement of chromodynamics is that each species of quark possess three aspects called colors, which have to do with the way it combines with other quarks to form a larger particle. The carriers of the color force—in the sense that photons are the carriers of electromagnetism—are called *gluons*. When a quark interacts with a gluon, it may change its color. The bizarre feature of chromodynamics is that none of its ingredients, neither quarks nor gluons nor color itself, has any real meaning outside the hadron. Quarks and gluons cannot be "shown to exist" in the way that other particles are. Baryons, made up of three quarks, contain one of each color, and so "appear" colorless.

Yet one more force is needed to describe the known world: the *weak force*. If the color force holds the nucleus together, the weak force may cause it to fall apart. It is a main feature of radioactive atomic nuclei. Nuclei are radioactive when they contain too many neutrons or protons. Such nuclei seek a state of equilibrium, which they achieve as some of their particles disintegrate, eject leptons and a quantity of energy, and change their identity. For example, a free neutron lives an average of 10 minutes before spitting out an electron and an antineutrino and changing into a proton.

The weak force is the force behind this process of decay, and like the forces of gravity, electromagnetism and color, it is essential to life. It causes the nuclear decay that produces the Sun's energy. It was necessary for the synthesis of chemical elements that took place before the earth was born. Without the weak force, the Sun and other stars would have shut down long ago and Spaceship Earth, if it existed at all, would be a cold and dreary place made up of pure hydrogen.

For many years, weak, electromagnetic, nuclear and gravitational forces seemed to be entirely different. Graduate students bought four textbooks, took four courses, and then decided in which force they would specialize. Everyone agreed that this was an unfortunate state of affairs and longed for a unified theory of all four forces. Einstein spent his last years trying to unify gravity and electromagnetism. He failed, as have all his successors.

However, progress is being made in the unification of all the forces except gravity. Recent theoretical advances and experimental results convincingly suggest that the weak force and electromagnetism are different facets of one unified theory. It is conjectured that the carriers of the weak force are observable particles called *intermediate vector bosons* with masses about 100 times greater than the proton's mass. Existing particle accelerators are not powerful enough to produce and detect these conjectured particles. The next generation of accelerators—if society sees fit to build them—will tell us whether or not this grand synthesis is correct.

If weak forces and electromagnetism are united, can the color-nuclear force be far behind? A unified theory of all forces except gravity seems almost at hand. And perhaps the day will come when we can deal with gravity, too, and establish the long-sought rapprochement between the celestial and the terrestrial worlds. Only then need we concern ourselves with Bertrand Russell's lament that physical science is "approaching the stage where it will be complete, and therefore uninteresting."

Nature is known to use three species of quarks (up, down, and strange) each of which comes in three colors. We are sure of the number of colors—both experiment and theory require that there be just three. Unfortunately, current theory is not powerful enough to predict the number of quark species. And experiment merely says that three are needed to describe known hadrons. There could, however, be other quark species that are constituents of hadrons not yet discovered.

Aesthetic arguments led J. D. Bjorken and me to conjecture a fourth quark more than a decade ago. Since leptons and quarks are most fundamental, and since there are four kinds of leptons, should there not also be four kinds of quarks? We called our construct the *charmed quark,* for we were fascinated and pleased by the symmetry it brought to the subnuclear world.

The case for *charm*—or the fourth quark—became much firmer when it was realized that there was a serious flaw in the familiar three-quark theory, which predicted that strange particles would sometimes decay in ways that they did not. In an almost magical way, the existence of the charmed quark prohibits these unwanted and unseen decays, and brings the theory into agreement with experiment. Thus did my recent collaborators, John Iliopoulos, Luciano Maiani, and I justify another definition of charm as a magical device to avert evil.

By the spring of 1974, many physicists were convinced of the necessary existence of charm, and of hadrons containing charmed quarks. Iliopoulos wagered several cases of fine wine on its imminent discovery; I offered to eat my hat if it were not experimentally confirmed within two years. Long articles were written to explain how best to search for charm.

The first great experimental revelation took place in November 1974 with the simultaneous discovery of a new hadron both at Brookhaven National Laboratory and at the Stanford Linear Accelerator Center. Today it is generally (but not unanimously) agreed that the new hadron is made up of one charmed quark and one charmed antiquark. The charm theory required that the new hadron be one member of a family of particles, and roughly predicted the properties of the other members. Experimenters responded in the spring of 1975 by discovering these predicted particles, thus further confirming the charm interpretation.

The original new hadron is variously called J or psi by its codiscoverers. Some theorists, pushing an analogy with positronium, which is an "atom" made up of an electron and a positron, call it *charmonium*. (Others, unable to contain their excitement, suggest "pandemonium".)

The discovery of the charmonium was an event of the utmost importance in elementary-particle physics. Nothing so exciting had happened in many years. For believers in quarks the new particle was the first experimental indication that a fourth quark existed. The successful interpretation of charmonium as a quark-antiquark combination, together with the difficulty in finding an attractive alternative hypothesis, led many doubters to see the error of their ways. As a result of the discovery of psi-J and its kin, the quark model has become orthodox philosophy.

According to the rules of the game, the charmed quark must be able to combine with ordinary quarks to form a new kind of matter, as well as with its own antiparticle to form charmonium. *Charmed hadrons* are those that contain just one charmed quark or antiquark, and experiments are now being performed or designed to find them.

Collisions of high-energy neutrinos with atomic nuclei should occasionally produce charmed particles, if the theory is correct. The charmed particles are not expected to be stable; they must decay by virtue of the weak force into other particles, leaving characteristic signals of their transient existence. If the collision is observed in a bubble chamber or a spark chamber, the decay products of the charmed particle can be identified. Indirect evidence of this kind for the existence of charmed particles has accumulated at laboratories in this country and abroad over the past two years. Something new and exciting was observed, but whether it was precisely charm remained uncertain until very recently.

It is in the collisions of high-energy electrons and positrons moving in opposite directions that physicists anticipated the copious production of charmed particles. The only existing laboratories where this can be done are SPEAR in Stanford, California, and DORIS, in Hamburg, Germany. On May 8, 1976 Gerson Goldhaber—one of a group of scientists working at SPEAR—telephoned to tell me that convincing evidence for a charmed particle had finally been found. The particle displayed exactly the properties that had been predicted for it. This information was given to me in confidence, for the experimenters were not yet prepared to announce their discovery. It was all but impossible for me to keep the secret. After all, John's wine and my hat had been saved in the nick of time. Experimental physicists should be kept busy for years to come finding other charmed particles and cataloguing their properties.

Philip Handler, president of the National Academy of Sciences, wrote just last month that in the past 30 years, "Man learned for the first time the nature of life, the structure of the cosmos, and the forces that shape the planet, although the interior of the nucleus became if anything, even more puzzling." This dismal assessment of my discipline is not quite up to date. With charm found, many seemingly unrelated pieces of the subnuclear puzzle have come together—quarks, color, unified theories—and we seem close to a new synthesis.

Having completed our tour of the subnuclear jungle, we must look to the future. For reasons of pedagogy and personal conviction, I have implied that a conceptually simple and empirically correct theory of the microworld is emerging. This remains to be seen.

Much theoretical analysis remains. Quark confinement by the color force seems too good an idea not to be true, but it must be proven to work. There remain urgent experimental questions, as well. Will the charmed particles behave in detail as the theory says they must? Are four quarks enough or will we need even more? Will the intermediate vector bosons be found?*

These, and other "technical questions" should be answered within a few years, and let us assume they will confirm the general picture we have sketched. Still, the story would be far from finished. True, hadrons are made of quarks just as nuclei are made of hadrons. We have descended yet one more level in the microworld. But even as few as four quarks, each in three colors—and at least four leptons to boot—seem like too many for them to be really basic.

* *Author's note:* These questions have been answered. Charmed particles behave as they should. There are six kinds of quarks, not four. The intermediate vector bosons have been found.

Recent theoretical developments have been exceedingly conservative, for they are based almost exclusively on the conceptual developments of the first third of this century: the quantum mechanics, and Einstein's special theory of relativity. Nothing done in physics since compares with the grandeur of these accomplishments. But we cannot seriously think that Nature has exhausted her bag of tricks. There will be such revolutions again. Just perhaps, the seeds to the next one are to be found in the tantalizing but incomplete theories of today.

Quarks with Color and Flavor

Atomos, the Greek root of "atom," means indivisible, and it was once thought that atoms were the ultimate, indivisible constituents of matter, that is, they were regarded as elementary particles. One of the principal achievements of physics in the twentieth century has been the revelation that the atom is not indivisible or elementary at all but has a complex structure. In 1911, Ernest Rutherford showed that the atom consists of a small, dense nucleus surrounded by a cloud of electrons. It was subsequently revealed that the nucleus itself can be broken down into discrete particles, protons and neutrons, and since then a great many related particles have been identified. During the past decade, it has become apparent that those particles too are complex, rather than elementary. They are now thought to be made up of the simpler things called quarks. A solitary quark has never been observed, in spite of many attempts to isolate one. Nonetheless, there are excellent grounds for believing they do exist. More important, quarks may be the last in the long series of progressively finer structures. They seem to be truly elementary.

When the quark hypothesis was first proposed, there were supposed to be three kinds of quark. The revised version of the theory I shall describe here requires 12 kinds. In the whimsical terminology that has evolved for the discussion of quarks, they are said to come in four flavors, and each flavor is said to come in three colors. ("Flavor" and "color" are, of course, arbitrary labels; they have no relation to the usual meanings of those words.) One of the quark flavors is distinguished by the property called charm (another arbitrary term). The concept of charm was suggested in 1964, but for a time it remained an untested conjecture.

Several experimental findings, including the discovery of particles called J, or psi, can be interpreted as supporting the charm hypothesis.

The basic notion that some subatomic particles are made of quarks has gained widespread acceptance, even in the absence of direct observational evidence. The more elaborate theory incorporating color and charm remains much more speculative. The views presented here are my own, and they are far from being accepted dogma. On the other hand, a growing body of evidence argues that these novel concepts must play some part in the description of nature. They help to bring together many seemingly unrelated theoretical developments to form an elegant picture of the structure of matter. Indeed, quarks are at once the most rewarding and the most mystifying creation of modern particle physics. They are remarkably successful in explaining the structure of subatomic particles, but we cannot yet understand why they should be so successful.

The particles thought to be made up of quarks form the class called the hadrons. They are the only particles that interact through the "strong" force. Included are protons and neutrons, and indeed it is the strong force that binds protons and neutrons together to form atomic nuclei. The strong force is also responsible for the rapid decay of many hadrons.

Another class of particles, defined in distinction to the hadrons, are the leptons. There are just four of them: the electron and the electron neutrino and the muon and the muon neutrino (and their four antiparticles). The leptons are not subject to the strong force. Because the electron and the muon bear an electric charge, they "feel" the electromagnetic force, which is roughly 100 times weaker than the strong force. The two kinds of neutrino, which have no electric charge, feel neither the strong force nor the electromagnetic force, but interact solely through a third kind of force, weaker by several orders of magnitude, called the weak force. The strong force, the electromagnetic force, and the weak force, together with gravitation, are believed to account for all interactions of matter.

Leptons give every indication of being elementary particles. The electron, for example, behaves as a point charge, and even when it is probed at the energies of the largest particle accelerators, no internal structure can be detected. Hadrons, on the other hand, seem complex. They have a measurable size: about 10^{-13} centimeter. Moreover, there are hundreds of them. Finally, all hadrons, with the significant exception of the proton and the antiproton, are unstable in isolation. They decay into stable particles, such as protons, electrons, neutrinos or photons. (The photon, which is the carrier of the electromagnetic force, is in a category apart; it is neither a lepton nor a hadron.)

Hadrons are subdivided into three families: baryons, antibaryons and mesons. The baryons include the proton and the neutron; the mesons include such particles as the pion. Baryons can be neither created nor destroyed, except as pairs of baryons and antibaryons. This principle defines a conservation law, and it can be treated most conveniently in the system of bookkeeping that assigns simple numerical values, called quantum numbers, to conserved properties. In this case, the quantum number is called baryon number. For baryons, it is $+1$, for antibaryons -1, and for mesons 0. The conservation of baryon number then reduces to the rule that in any interaction the sum of the baryon numbers cannot change.

Baryon number provides a means of distinguishing baryons from mesons, but it is an artificial means, and it tells us nothing about the properties of the two kinds of particle. A more meaningful distinction can be established by examining another quantum number, spin angular momentum.

Under the rules of quantum mechanics, a particle, or a system of particles, can assume only certain specified states of rotation, and hence can have only discrete values of angular momentum. The angular momentum is measured in units of $h/2\pi$, where h is Planck's constant, equal to about 6.6×10^{-27} erg second. Baryons are particles with a spin angular momentum measured in half-integral units, that is, with values of half an odd integer, such as ½ or 3⁄2. Mesons have integral values of spin angular momentum, such as 0 or 1.

The difference in spin angular momentum has important consequences for the behavior of the two kinds of hadron. Particles with integral spin are said to obey Bose-Einstein statistics (and are therefore called bosons). Those with half-integral spin obey Fermi-Dirac statistics (and are called fermions). In this context "statistics" refers to the behavior of a population of identical particles. Those that obey Bose-Einstein statistics can be brought together without restriction; an unlimited number of pions, for example, can occupy the same state. The Fermi-Dirac statistics, on the other hand, require that no two particles within a given system have the same energy and be identical in all their quantum numbers. This statement is equivalent to the exclusion principle, formulated in 1925 by

Wolfgang Pauli. He applied it, in particular, to electrons, which have a spin of ½ and are therefore fermions. It requires that each energy level in an atom contain only two electrons, with their spins aligned in opposite directions.

One of the clues to the complex nature of the hadrons is that there are so many of them. Much of the endeavor to understand them has consisted of a search for some ordering principle that would make sense of the multitude.

The hadrons were first organized into small families of particles called charge multiplets or isotopic-spin multiplets; each multiplet consists of particles that have approximately the same mass and are identical in all their other properties except electric charge. The multiplets have one, two, three or four members. The proton and the neutron compose a multiplet of two (a doublet); both are considered to be manifestations of a single state of matter, the nucleon, with an average mass equivalent to an energy of .939 GeV (billion electron volts). The pion is a triplet with an average mass of .137 GeV and three charge states: +1, 0 and −1. In strong interactions, the members of a multiplet are all equivalent since electric charge plays no role in strong interactions.

In 1962, a grander order was revealed when the charge multiplets were organized into "supermultiplets" that revealed relations between particles that differ in other properties, in addition to charge. The creation of the supermultiplets was proposed independently by Murray Gell-Mann of the California Institute of Technology and by Yuval Ne'eman of Tel-Aviv University. The introduction of the new system led directly to the quark hypothesis.

The grouping of hadrons into supermultiplets involves eight quantum numbers and has been referred to as the "eightfold way." Its mathematical basis is a branch of group theory invented in the nineteenth century by the Norwegian mathematician Sophus Lie. The Lie group that generates the eightfold way is called SU(3), which stands for the special unitary group of matrices of size 3×3. The theory requires that all hadrons belong to families corresponding to representations of the group SU(3). The families can have one, three, six, eight, ten or more members. If the eightfold way were an exact theory, all the members of a given family would have the same mass. The eightfold way is only an approximation, however, and within the families there are significant differences in mass.

The construction of the eightfold way begins with the classification of the hadrons into broad families sharing a common value of spin angular momentum. Each family of particles with identical spin is then depicted by plotting the distribution of two more quantum numbers: isotopic spin and strangeness.

Isotopic spin has nothing to do with the spin of a particle; it was given its name because it shares certain algebraic properties with the spin quantum number. It is a measure of the number of particles in a multiplet, and it is calculated according to the formula that the number of particles in the multiplet is one more than twice the isotopic spin. Thus the nucleon (a doublet) has an isotopic spin of ½; for the pion triplet the isotopic spin is 1.

Strangeness is a quantum number introduced to describe certain hadrons first observed in the 1950s and called strange particles because of their anomalously long lifetimes. They generally decay in from 10^{-10} to 10^{-7} second. Although that is a brief interval by everyday standards, it is much longer than the lifetime of 10^{-23} second, characteristic of many other hadrons.

Like isotopic spin, strangeness depends on the properties of the multiplet, but it measures the distribution of charge among the particles, rather than their number. Strangeness quantum number is equal to twice the average charge (the sum of the charges divided by the number of particles in a multiplet) minus the baryon number. By this contrivance it is made to vanish for all hadrons except the strange ones. The triplet of pions, for example, has an average charge of 0 and a baryon number of 0; its strangeness is therefore also 0. The nucleon doublet has an average charge of $+½$ and a baryon number of $+1$, so that those particles too have a strangeness of 0. On the other hand, the lambda particle is a neutral baryon that forms a family of one (a singlet). Its average charge of 0 and its baryon number of $+1$ give it a strangeness of -1.

On a graph that plots electric charge against strangeness, hadrons form orderly arrays. Mesons with a spin angular momentum of 0 compose an octet and a singlet; the octet is represented graphically as a hexagon with a particle at each vertex and two particles in the center, and the singlet is represented as a point at the origin. The mesons with a spin of 1 form an identical representation, and so do the baryons with a spin of ½. Finally, the baryons with a spin of 3⁄2 form a decimet (a group of 10) that can be graphed as a large triangle made up of a singlet, a doublet, a triplet and a quartet. The eightfold way was initially greeted with some skepticism, but the discovery in 1964 of the negatively charged omega particle, the predicted singlet in the baryon decimet, made converts of us all.

The regularity and economy of the supermultiplets are aesthetically satisfying, but they are also somewhat mystifying. The known hadrons do fit into such families, without exception. Mesons come only in families of one and eight, and baryons come only in families of one, eight, and ten. The singlet, octet and decimet, however, are only a few of many possible representations of SU(3). Families of three particles or six particles are entirely plausible, but they are not observed. Indeed, the variety of possible families is in principle infinite. Why, then, do only three representations appear in nature? It early became apparent that the eightfold way is, in some approximate sense, true, but it was also plain from the start that there is more to the story.

In 1963, an explanation was proposed, independently by Gell-Mann and by George Zweig, also of the California Institute of Technology. They perceived that the unexpected regularities could be understood if all hadrons were constructed from more fundamental constituents, which Gell-Mann named quarks. The quarks were to belong to the simplest, nontrivial family of the eightfold way: a family of three. (There is also, of course, another family of three antiquarks.)

The quarks are required to have rather peculiar properties. Principal among these is their electric charge. All observed particles, without exception, bear integer multiples of the electron's charge; quarks, however, must have charges that are fractions of the electron's charge. Gell-Mann designated the three quarks u, d and s, for the arbitrary labels "up," "down" and "sideways."

The mechanics of the original quark model are completely specified by three simple rules. Mesons are invariably made of one quark and one antiquark. Baryons are invariably made of three quarks and antibaryons of three antiquarks. No other assemblage of quarks can exist as a hadron. The combinations of the three quarks under these rules are sufficient to account for all the hadrons that had been observed or predicted at the time. Furthermore, every allowed combination of quarks yields a known particle.

Many of the necessary properties of the quarks can be deduced from these rules. It is mandatory, for example, that each of the quarks be assigned a baryon number of $+\frac{1}{3}$ and each of the antiquarks a baryon number of $-\frac{1}{3}$. In that way, any aggregate of three quarks has a total baryon number of $+1$ and hence defines a baryon; three antiquarks yield a particle with a baryon number of -1, an antibaryon. For mesons the baryon numbers of the quarks ($+\frac{1}{3}$ and $-\frac{1}{3}$) cancel, so that the meson, as required, has a baryon number of 0.

In a similar way, the angular momentum of hadrons is described by giving quarks half-integral units of spin. A particle made of an odd number of quarks, such as a baryon, must therefore also have half-integral spin, conforming to the known characteristics of baryons. A particle made of an even number of quarks, such as a meson, must have integral spin.

The u quark and the s quark compose an isotopic-spin doublet: they have nearly the same mass and they are identical in all other properties, except electric charge. The u quark is assigned a charge of $+\frac{2}{3}$ and the d quark is assigned a charge of $-\frac{1}{3}$. The average charge of the doublet is therefore $+\frac{1}{6}$ and twice the average charge is $+\frac{1}{3}$; since the baryon number of all quarks is $+\frac{1}{3}$, the definition of strangeness gives both the u and the d quarks a strangeness of 0. The s quark has a larger mass than either the u or the d and makes up an isotopic-spin singlet. It is given an electric charge of $-\frac{1}{3}$ and consequently has a strangeness of -1. The antiquarks, denoted by writing the quark symbol with a bar over it, have opposite properties. The $\bar{u}$ has a charge of $-\frac{2}{3}$ and the $\bar{d}$ $+\frac{1}{3}$; both have zero strangeness. The $\bar{s}$ antiquark has a charge of $+\frac{1}{3}$ and a strangeness of $+1$.

Just two of the quarks, the u and the d, suffice to explain the structure of all hadrons encountered in ordinary matter. The proton, for example, can be described by assembling two u quarks and a d quark; its composition is written uud. A quick accounting will show that all the properties of the proton determined by its quark constitution are in accord with the measured values. Its charge is equal to $\frac{2}{3} + \frac{2}{3} - \frac{1}{3}$, or $+1$. Similarly, its baryon number can be shown to be $+1$ and its spin $\frac{1}{2}$. A positive pion is composed of a u quark and a $\bar{d}$ antiquark (written $u\bar{d}$). Its charge is $\frac{2}{3} + \frac{1}{3}$, or $+1$; its spin and baryon number are both 0.

The third quark, s, is needed only to construct strange particles, and indeed it provides an explicit definition of strangeness: A strange particle is one that contains at least one s quark or $\bar{s}$ antiquark. The lambda baryon, for example, can be shown from the charge distribution of its multiplet to have a strangeness of −1; that result is confirmed by its quark constitution of uds. Similarly, the neutral K meson, a strange particle, has a strangeness of +1, as confirmed by its composition of $d\bar{s}$.

Until quite recently, these three kinds of quark were sufficient to describe all known hadrons. But, experiments show that certain hadrons cannot be explained in terms of the original three quarks. The experiments imply the existence of a fourth kind of quark, called the charmed quark, designated c.

The statement that the u, d and s quarks are sufficient to construct all the observed hadrons can be made more precisely in the mathematical formalism of the eightfold way. Since a meson is made up of one quark and one antiquark, and since there are three kinds, or flavors, of quark, there are nine possible combinations of quarks and antiquarks that can form a meson. It can be shown that one of these combinations represents a singlet and the remaining eight form an octet. Similarly, since a baryon is made up of three quarks, there are 27 possible combinations of quarks that can make up a baryon. They can be broken up into a singlet, two octets and a decimet. Those groupings correspond exactly to the observed families of hadrons. The quark theory thus explains why only a few of the possible representations of SU(3) are realized in nature as hadron supermultiplets.

Quarks rules provide a remarkably economical explanation for the formation of the observed hadron families. What principles, however, can explain quark rules, which seem quite arbitrary? Why is it possible to bind together three quarks but not two or four? Why can we not create a single quark in isolation? A line of thought that leads to possible answers to these questions appeared at first as a defect in the quark theory.

As we have seen, it is necessary that quarks have half-integral values of spin angular momentum; otherwise the known spins of baryons and mesons would be predicted wrongly. Particles with half-integral spin are expected to obey Fermi-Dirac statistics and are therefore subject to the Pauli exclusion principle: No two particles within a particular system can have exactly the same quantum numbers. Quarks, however, seem to violate the principle. In making up a baryon, it is often necessary that two identical quarks occupy the same state. The omega particle, for example, is made up of three s quarks, and all three must be in precisely the same state. That is possible only for particles that obey Bose-Einstein statistics.

We are at an impasse: quarks must have half-integral spin, but they must satisfy the statistics appropriate to particles having integral spin.

The connection between spin and statistics is an unshakable tenet of relativistic quantum mechanics. It can be deduced directly from the theory, and a violation has never been discovered. Since it holds for all other known particles, quarks could not reasonably be excluded from its dominion.

The concept that has proved essential to the solution of the quark statistics problem was proposed in 1964 by Oscar W. Greenberg of the University of Maryland. He suggested that each flavor of quark comes in three varieties, identical in mass, spin, electric charge and all other measurable quantities but different in an additional property, which has come to be known as color. The exclusion principle could then be satisfied, and quarks could remain fermions, because quarks in a baryon would not all occupy the same state. Quarks could differ in color, even if they were the same in all other respects.

The color hypothesis requires two additional quark rules. The first simply restates the condition that color was introduced to satisfy: Baryons must be made up of three quarks, all of which have different colors. The second describes the application of color to mesons: Mesons are made of a quark and an antiquark of the same color, but with equal representation of each of the three colors. The effect of these rules is that no hadron can exhibit net color. A baryon invariably contains quarks of each of the three colors, say red, yellow and blue. In the meson one can imagine quark and antiquark as being a single color at any given moment, but continually and simultaneously changing color, so that over any measurable interval they will both spend equal amounts of time as red, blue and yellow quarks.

The price of the color hypothesis is a tripling of the number of quarks; there must be nine instead of three (with charm yet to be considered). At first it may also appear that we have greatly increased the number of hadrons, but that is an illusion. With color there seem to be nine times as many mesons and 27 times as many baryons, but the rules for assembling hadrons from colored quarks ensure that none of the additional particles are observable.

Although the quark rules imply that we will never see a colored particle, the color hypothesis is not merely a formal construct without predictive value. The increase it requires in the number of quarks can be detected in at least two ways. One is through the effect of color on the lifetime of the neutral pion, which almost always decays into two photons. Stephen L. Adler of the Institute for Advanced Study has shown that its rate of decay depends on the square of the number of quark colors. Just the observed lifetime is obtained by assuming that there are three colors.

Another effect of color can be detected in experiments in which electrons and their antiparticles, positrons, annihilate each other at high energy. The outcome of such an event is sometimes a group of hadrons and sometimes a muon and an antimuon. At sufficiently high energy, the ratio of the number of hadrons to the number of muon-antimuon pairs is expected to approach a constant value, equal to the sum of the squares of the charges of the quarks. Tripling the number of quarks also triples the expected value of the ratio. The experimental result at energies of from 2 GeV to 3 GeV is in reasonable agreement with the color hypothesis (which predicts a value of 2) and is quite incompatible with the original theory of quarks without color.

The introduction of the color quantum number solves the problem of quark statistics, but it once again requires a set of rules that seem arbitrary. The rules can be accounted for, however, by establishing another hypothetical symmetry group, analogous to the SU(3) symmetry proposed by Gell-Mann and by Ne'eman. The earlier SU(3) is concerned entirely with combinations of the three quark flavors; the new one deals exclusively with the three quark colors. Moreover, unlike the earlier theory, which is only approximate, color SU(3) is supposed to be an exact symmmetry, so that quarks of the same flavor but different color will have identical masses.

In the color SU(3) theory all the quark rules can be explained if we accept one postulate: All hadrons must be represented by color singlets; no larger multiplets can be allowed. A color singlet can be constructed in two ways: by combining an identically colored quark and antiquark with all three colors equally represented, or by combining three quarks or three antiquarks in such a way that the three colors are all included. These conditions, of course, are equivalent to the rules for building mesons, baryons and antibaryons, and they ensure that all hadrons will be colorless. There are no other ways to make a singlet in color SU(3); a particle made any other way would be a member of a larger multiplet, and it would display a particular color.

Although the color SU(3) theory of the hadrons can explain the quark rules, it cannot entirely eliminate the arbitrary element in their nature. We can ask a still more fundamental question: What explains the postulate that all hadrons must be color singlets? One approach to an answer, admittedly a speculative one, has been suggested by many investigators; it incorporates the color SU(3) model of the hadrons into one of the class of theories called gauge theories.

The color gauge theory postulates the existence of eight massless particles, sometimes called gluons, that are the carriers of the strong force, just as the photon is the carrier of the electromagnetic force. Like the photon, they are electrically neutral, and they have a spin of 1; they are therefore called vector bosons (bosons because they have integer spin and obey Bose-Einstein statistics, vector because a particle with a spin of 1 is described by a wave function that takes the form of a four-dimensional vector). Gluons, like quarks, have not been detected.

When a quark emits or absorbs a gluon, the quark changes its color but not its flavor. For example, the emission of a gluon might transform a red u quark into a blue or a yellow u quark, but it could not change it into a d or an s quark of any color. Since the color gluons are the quanta of the strong force, it follows that color is the aspect of quarks that is most important in the strong interactions. In fact, when describing interactions that involve only the strong force, one can virtually ignore the flavors of quarks.

The color gauge theory proposes that the force that binds together colored quarks represents the true character of the strong interaction. The more familiar strong interactions of hadrons (such as the binding of protons and neutrons in a nucleus) are manifestations of the same fundamental force, but the interactions of colorless hadrons are no more than a pale remnant of the underlying interaction between colored quarks. Just as the van der Waals force between molecules is only a feeble vestige of the electromagnetic force that binds electrons to nuclei, the strong force observed between hadrons is only a vestige of that operating within the individual hadron.

From these theoretical arguments one can derive an intriguing, if speculative, explanation of the confinement of quarks. It has been formulated by John Kogut and Kenneth Wilson of Cornell University and by Leonard Susskind of Yeshiva University. If it should be proved correct, it would show that the failure to observe colored particles (such as isolated quarks and gluons) is not the result of any experimental deficiency but is a direct consequence of the nature of the strong force.

The electromagnetic force between two charged particles is described by Coulomb's law: The force decreases as the square of the distance between the charges. Gravitation obeys a fundamentally similar law. At large distances both forces dwindle to insignificance. Kogut, Wilson and Susskind argue that the strong force between two colored quarks behaves quite differently: it does not diminish with distance, but remains constant, independent of the separation of the quarks. If their argument is sound, an enormous amount of energy would be required to isolate a quark.

Separating an electron from the valence shell of an atom requires a few electron volts. Splitting an atomic nucleus requires a few million electron volts. In contrast to these values, the separation of a single quark by just an inch from the proton of which it is a constituent would require the investment of 10^{13} GeV, enough energy to separate the author from the earth by some 30 feet. Long before such an energy level could be attained, another process would intervene. From the energy supplied in the effort to extract a single quark, a new quark and antiquark would materialize. The new quark would replace the one removed from the proton, and would reconstitute that particle. The new antiquark would adhere to the dislodged quark, making a meson. Instead of isolating a colored quark, all that is accomplished is the creation of a colorless meson. By this mechanism we are prohibited from ever seeing a solitary quark or a gluon or any combination of quarks or gluons that exhibits color.

If this interpretation of quark confinement is correct, it suggests an ingenious way to terminate the apparently infinite regression of finer structures in matter. Atoms can be analyzed into electrons and nuclei, nuclei into protons and neutrons, and protons and neutrons into quarks, but the theory of quark confinement suggests that the series stops there. It is difficult to imagine how a particle could have an internal structure if the particle cannot even be created.

Quarks of the same flavor, but different color, are expected to be identical in all properties except color; indeed, that is why the concept of color was introduced. Quarks that differ in flavor, however, have quite different properties. It is because the u quark and the d quark differ in electric charge that the proton is charged and the neutron is not. Similarly, it is because the s quark is considerably more massive than either the u or the d quark that strange particles are generally the heaviest members of their families. The charmed quark, c, must be heavier still, and charmed particles as a rule should therefore be heavier than all others. It is the flavor of quarks that brings variety to the world of hadrons, not their color.

As we have seen, flavors of quarks are unaffected by strong interactions. In a weak interaction, on the other hand, a quark can change its flavor (but not its color). Weak interactions also couple quarks to leptons. The classical example of this coupling is nuclear beta decay, in which a neutron is converted into a proton with the emission of an electron and an antineutrino. In terms of quarks, the reaction represents the conversion of a d quark to a u quark, accompanied by the emission of the two leptons.

Weak interactions are thought to be mediated by vector bosons, just as strong and electromagnetic interactions are. The principal one, labeled W, and long called the intermediate vector boson, was predicted in 1938 by Hideki Yukawa. It has an electric charge of -1, and it differs from the photon and the color gluons in that it has mass, indeed a quite large mass. Quarks can change their flavor by emitting or absorbing a W particle. Beta decay, for example, is interpreted as the emission of a W by a d quark, which converts the quark into a u; the W then decays to yield the electron and antineutrino. From this process, it follows that the W can also interact with leptons, and it thus provides a link between the two groups of apparently elementary particles.

The realization that the strong, weak and electromagnetic forces are all carried by the same kind of particle—bosons with a spin of 1—invites speculation that all three might have a common basis in some simple, unified theory. A step toward such a unification would be the reconciliation of the weak interactions and electromagnetism. Julian Schwinger of Harvard University attempted such a unification in the mid-1950s (when I was one of his doctoral students, working on these very questions). His theory had serious flaws. One was eliminated in 1961, when I introduced a second, neutral vector boson, now called Z, to complement the electrically charged W. Other difficulties persisted for 10 years, until in 1967, Steven Weinberg of Harvard and Abdus Salam of the International Center for Theoretical Physics in Trieste, independently suggested a resolution. By 1971 it was generally agreed, largely because of the work of Gerhard 't Hooft of the University of Utrecht, that the Weinberg-Salam conjecture is successful.

Through the unified weak and electromagnetic interactions, quarks and leptons are intimately related. These interactions "see" the four leptons and distinguish between the three quark flavors. The W particle can induce one kind of neutrino to become an electron and the other kind of neutrino to become a muon. Similarly, the W can convert a u quark into a d quark; it can also influence the u quark to become an s quark, although much less readily.

There is an obvious lack of symmetry in these relations. Leptons consist of two couples, married to each other by the weak interaction: the electron with the electron neutrino and the muon with the muon neutrino. Quarks, on the other hand, come in only three flavors, and so one must remain unwed. The scheme could be made much tidier if there were a fourth quark flavor, in order to provide a partner for the unwed quark. Both quarks and leptons would then consist of two pairs of particles, and each member of a pair could change into the other member of the same pair, simply by emitting a W. The desirability of such lepton-quark symmetry led James Bjorken and me, among others, to postulate the existence of a fourth quark in 1964. Bjorken and I called it the charmed quark. When provisions are made for quark colors, charm becomes a fourth quark flavor, and a new triplet of colored quarks is required. There are thus a total of 12 quarks.

Since 1964, several additional arguments for charm have developed. To me the most compelling of them is the need to explain the suppression of certain interactions, called strangeness-changing neutral currents. An explanation that relies on the properties of the charmed quark was presented in 1967 by John Iliopoulos, Luciano Maiani and me.

Strangeness-changing neutral currents are weak interactions in which the net electric charge of the hadrons does not change but the strangeness does; typically an s quark is transformed into a d quark, and two leptons are emitted. An example is the decay of the neutral K meson (a strange particle) into two oppositely charged muons. Such processes are found by experiment to be extremely rare. The three-quark theory cannot account for their suppression, and in fact the unified theory of weak and electromagnetic interactions predicts rates more than a million times greater than those observed.

The addition of a fourth quark flavor with the same electric charge as the u quark neatly accounts for the suppression, although the mechanism by which it does so may seem bizarre. With two pairs of quarks there are two possible paths for the strangeness-changing interactions, instead of just one when there are only three quarks. In the macroscopic world, the addition of a second path, or channel, would be expected always to bring an increase in the reaction rate. In a world governed by quantum mechanics, however, it is possible to subtract as well as to add. As it happens, a sign in the equation that defines one of the reactions is negative, and the two interactions cancel each other.

The addition of a fourth quark flavor must obviously increase the number of hadrons. In order to accommodate the newly predicted particles in supermultiplets, the eightfold way must be expanded. In particular, another dimension must be added to the graphs employed to represent the families, so that the plane figures of the earlier symmetry become Platonic and Archimedean solids.

To the meson octet are added six charmed particles and one uncharmed particle to make up a new family of 15. It is represented as a cuboctahedron, in which one plane contains the hexagon of the original uncharmed meson octet. Baryon octets and decimet are expected to form two families having 20 members each. They are represented as a tetrahedron truncated at each vertex and as a regular tetrahedron. In addition, there is a smaller regular tetrahedron consisting of just four baryons. Again, each figure contains one plane of uncharmed particles (see Figure 1).

It now appears that the first of the new particles to be discovered is a meson that is not itself charmed. That conclusion is based on the assumption that the predicted meson is the same particle as the J or psi particle. The announcement of the discovery was made simultaneously by Samuel C. C. Ting and his colleagues at the Brookhaven National Laboratory and by Burton Richter, Jr., and a group of other physicists at the Stanford Linear Accelerator Center (SLAC). At Brookhaven it was named J, at Stanford psi. Here I shall adopt the name J. For two excited states of the same particle, however, the names psi′ and psi″ will be employed, since they were seen only in the SLAC experiments.

The J particle was found as a resonance, an enhancement at a particular energy in the probability of an interaction between other particles. At Brookhaven the resonance was detected in the number of electron-positron pairs produced in collisions between protons and atomic nuclei. At SLAC it was observed in the products of annihilations of electrons and positrons. The energy at which the resonances were observed—and thus the energy or mass of the J particle—is about 3.1 GeV.

The J particle decays in about 10^{-20} second, certainly a brief interval, but nevertheless a 1,000 times longer than the expected lifetime of a particle having the J's mass. The considerable excitement generated by the discovery of the J was largely a result of its long lifetime.

A great many explanations of the particle were proposed; for example, it was suggested that it might be the Z. I believe there is good reason to interpret the J as being a meson made up of a charmed quark and a charmed antiquark, that is, a meson with the quark constitution $c\bar{c}$. Thomas Appelquist and H. David Politzer of Harvard have named such a meson "charmonium," by analogy to positronium, a bound state of an electron and a positron. Charmonium is without charm because the charm quantum numbers of its quarks ($+1$ and -1) add up to zero.

The charmonium hypothesis can account for the anomalous lifetime of the J, if one considers the ultimate fate of the decaying particle's quarks. There are three possibilities: they can be split up to become constituents of two daughter hadrons, they can both become part of a single daughter particle or they can be annihilated. An empirical rule, first noted by Zweig, states that decays of the first kind are allowed but the other two are suppressed. For the J particle to decay in the allowed manner, it must create two charmed particles, that is, two hadrons, one containing a charmed quark and the other a charmed antiquark. That decay is possible only if the mass of the J is greater than the combined masses of the charmed daughter particles. There is reason to believe the lightest charmed particle has a mass greater than half of the mass of the J, and it therefore appears that the J cannot decay in the allowed mode. The J cannot decay in the second way, either, keeping both its quarks in a single particle, because the J is the least massive state containing a charmed quark and a charmed antiquark. It must therefore decay by the annihilation of its quarks, a decay suppressed by Zweig's rule. The

FIGURE 1 (*right*). *Supermultiplets of hadrons that include the predicted charmed particles can be arranged as polyhedrons. Each supermultiplet consists of particles with the same value of spin angular momentum. Within each supermultiplet the particles are assigned positions according to three quantum numbers: positions on the dark shaded planes are determined by isotopic spin and strangeness; the planes themselves indicate values of charm. The mesons are represented by a point (a) and by an Archimedean solid called a cuboctahedron (b), which comprises 15 particles, including six charmed ones. The mesons shown are those with a spin of 1, but all mesons fit the same point and cuboctahedron representations. The baryons form a small regular tetrahedron (c) of four particles, a truncated tetrahedron (d) of 20 particles and a larger regular tetrahedron (e), also made up of 20 particles. Both mesons and baryons are identified by their quark constitution, and for those particles that have been observed the established symbol is also given. Each figure contains one plane, shown by light shaded areas, of uncharmed particles that are identical with earlier representations of the "eightfold way."*

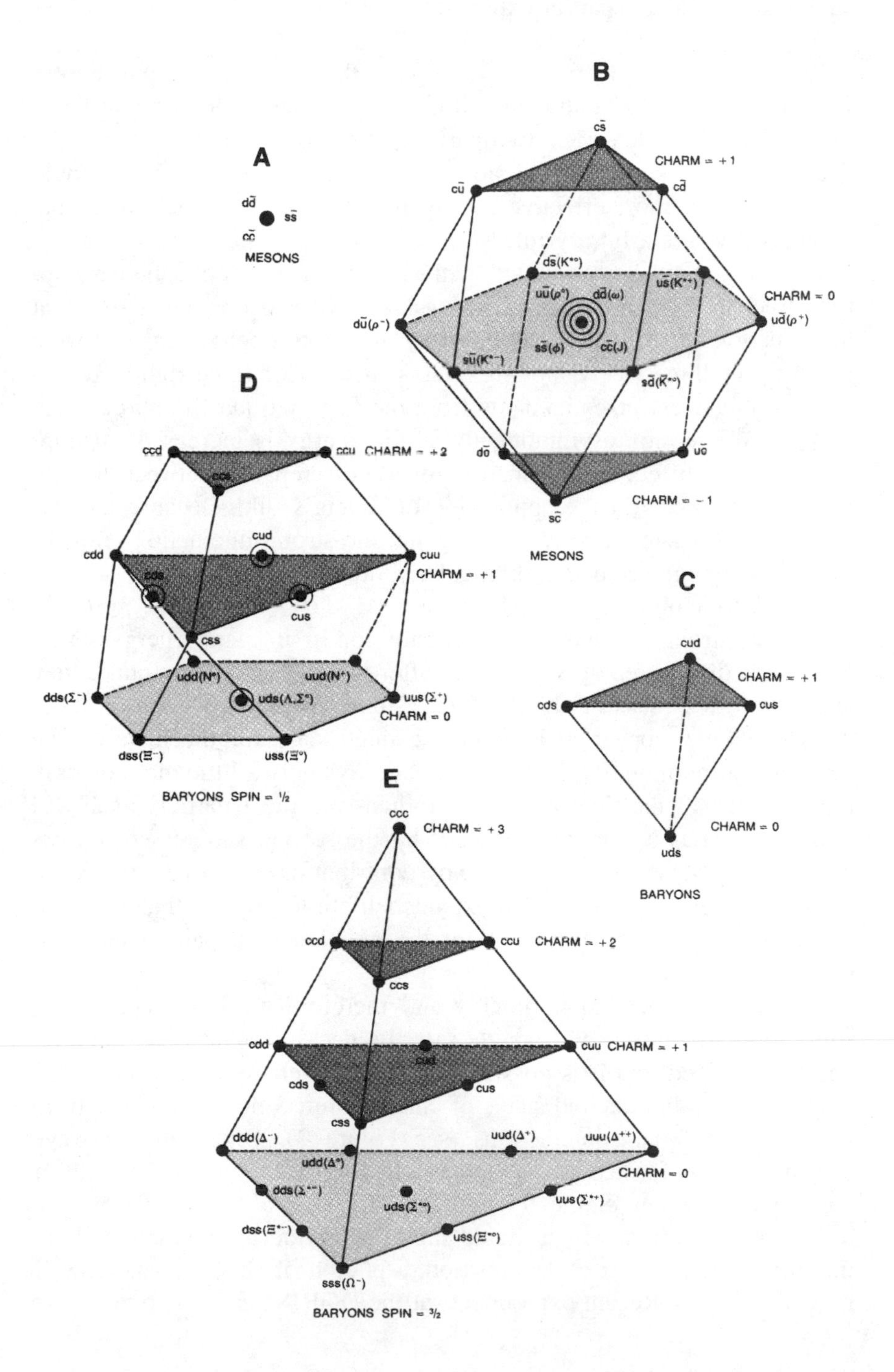
A
dd̄
ss̄
cc̄
MESONS
B
cs̄
CHARM = +1
cū
cd̄
ds̄(K*°)
us̄(K*+)
uū(ρ°)
dd̄(ω)
CHARM = 0
dū(ρ⁻)
ud̄(ρ⁺)
ss̄(φ)
cc̄(J)
sū(K*⁻)
sd̄(K̄*°)
dc̄
uc̄
CHARM = −1
sc̄
MESONS
C
cud
CHARM = +1
cds
cus
CHARM = 0
uds
BARYONS
D
ccd
ccu CHARM = +2
ccs
cud
cdd
cuu
CHARM = +1
cds
cus
css
udd(N°)
uud(N⁺)
dds(Σ⁻)
uds(Λ,Σ°)
uus(Σ⁺)
CHARM = 0
dss(Ξ⁻)
uss(Ξ°)
BARYONS SPIN = ½
E
ccc
CHARM = +3
ccd
ccu
CHARM = +2
ccs
cdd
cuu CHARM = +1
cud
cds
cus
css
ddd(Δ⁻)
uud(Δ⁺)
uuu(Δ⁺⁺)
udd(Δ°)
CHARM = 0
dds(Σ*⁻)
uds(Σ*°)
uus(Σ*⁺)
dss(Ξ*⁻)
uss(Ξ*°)
sss(Ω⁻)
BARYONS SPIN = 3/2

suppression offers a partial explanation for the particle's extended lifetime.

Zweig's rule was formulated to explain the decay of the phi meson, which is made up of a strange quark and a strange antiquark and has a mass of about 1 GeV. The two particles are closely analogous, but the decay of the J is appreciably slower than that of the phi. Why should Zweig's rule be more effective for J than it is for phi? Furthermore, what explains Zweig's arbitrary rule?

A possible answer is provided by the theoretical concept called asymptotic freedom, which holds that strong interactions become less strong at high energy. At sufficiently high energy, the proton behaves as if it were made up of three freely moving quarks, instead of three tightly bound ones. The concept takes its name from the fact that quarks approach the state of free motion asymptotically as the energy is increased. Asymptotic freedom offers an explanation for the discrepancy between the phi and the J particles in the application of Zweig's rule. Because the J is so massive, or alternatively so energetic, the strong interaction is of diminished strength, and it is particularly difficult for the quark and the antiquark to annihilate each other.

Like positronium, charmonium should appear in many energy states. Two were discovered at SLAC soon after the first state was found; they are psi′, with a mass of about 3.7 GeV, and psi″, with a mass of about 4.1 GeV. They appear to be simple excited states of the lowest-lying state of charmonium, the J particle. Psi′ decays only a little more quickly than J, and half the time its decay products are the J particle itself and two pions. Thus it sometimes decays by the second suppressed process described by Zweig's rule, that is, by contributing both of its quarks to a single daughter particle. The extended lifetime implies that psi′ also lies below the energy threshold for the creation of a pair of charmed particles.

Psi″ decays much more quickly and therefore must be decaying in some mode permitted by Zweig's rule. Its decay products have not yet been determined, but it is possible they include charmed hadrons.

Numerous other excited states of charmonium follow inevitably from the theory of quark interactions (see Figure 2). One, called *p*-wave charmonium, is formed when the particle takes on an additional unit of angular momentum. Some fraction of the time psi′ should decay into *p*-wave charmonium, which should subsequently decay predominantly to the ground state, J. At each transition, a photon of characteristic energy must be emitted. Recent experiments at the DORIS particle-storage rings

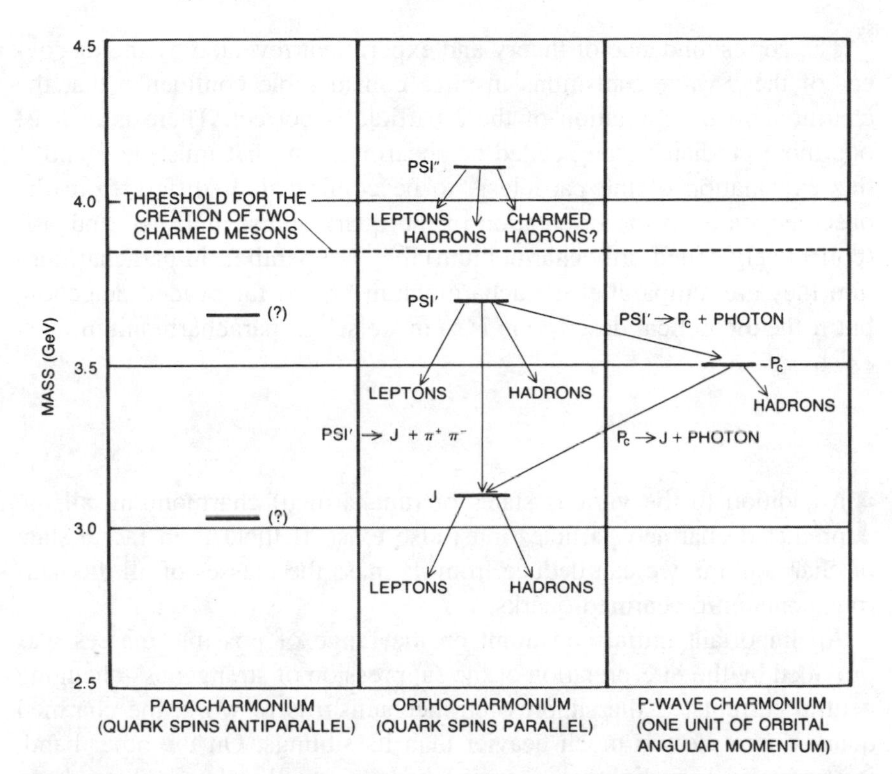

FIGURE 2. *Charmonium must exist at several energy levels, distinguished by the state of motion of constituent quarks. The J particle is the ground state of orthocharmonium, in which the quark spins are parallel. Two excited states of orthocharmonium, designated psi′ and psi″, were discovered at SLAC shortly after the J particle. Psi′, like the J particle, seems to be too light to decay into two charmed hadrons, but the rapid decay of psi″ suggests that it has the necessary mass. Two other forms of the particle, called paracharmonium, in which the quark spins are antiparallel, have not been discovered.* P-*wave charmonium, in which the quarks have a unit of orbital angular momentum in addition to spin angular momentum, may have been detected at the German Electron Synchrotron in Hamburg and at SLAC.*

of the German Electron Synchrotron in Hamburg have apparently detected the decays associated with the *p*-wave particle. In a few percent of its decays psi′ yields the J particle and two photons, with energies of .2 GeV and .4 GeV. At SLAC, psi′ has been found to decay into an intermediate state and a single photon with an energy of .2 GeV. The intermediate state, which is presumably the same particle as the one observed at DORIS, then decays directly into hadrons.

The correspondence of theory and experiment revealed by the discovery of the *p*-wave transitions inspires considerable confidence that the charmonium interpretation of the J particle is correct. There is at least one more predicted state, called paracharmonium, that must be found if this explanation of the particle is to be confirmed. It differs from the observed states in the orientation of the quark spins: in J, psi′ and psi″ (collectively called orthocharmonium) they are parallel; in paracharmonium they are antiparallel. Paracharmonium has so far evaded detection, but if the theoretical description is to make sense, paracharmonium must exist.

In addition to the various states of (uncharmed) charmonium, all the predicted charmed particles must also exist. If the J is in fact a state of charmonium, we can deduce from its mass the masses of all the hadrons containing charmed quarks.

An important initial constraint on the range of possible masses was provided by the interpretation of the suppression of strangeness-changing neutral currents. If the suppression mechanism is to work, the charmed quark cannot be too much heavier than its siblings. On the other hand, it cannot be very light or charmed hadrons would already have been observed. An estimate from these conditions suggested that charmed particles would be found to have masses of about 2 or 3 GeV.

After the discovery of the J, I performed a more formal analysis with my colleagues at Harvard, Alvaro De Rújula and Howard Georgi. So did many others. Our estimates indicate that the least massive charmed states are mesons made up of a *c* quark and a $\bar{u}$ or $\bar{d}$ antiquark; their mass should fall between 1.8 GeV and 2.0 GeV. A value within that range could be in agreement with the supposition that psi′ lies below the threshold for the creation of a pair of charmed mesons, but psi″ lies above it.

The least massive charmed baryon has a quark composition of udc; we predict that its mass is near 2.2 GeV. As might be expected, since the c quark is the heaviest of the four, the most massive predicted charmed hadron is the ccc baryon. We estimate its mass at about 5 GeV.

An important principle guiding experimental searches for charmed hadrons is the requirement that in most kinds of interaction charmed particles can be created only in pairs. Two hadrons must be produced, one containing a charmed quark, the other a charmed antiquark; the obvious consequence is a doubling of the energy required to create a charmed particle. An important exception to this rule is the interaction of neutrinos with other kinds of particles, such as protons. Neutrino events are exempt because neutrinos have only weak interactions and quark flavor can be changed in weak processes. Many experimental techniques have been tried in the search for charm, yet no charmed particle has been unambiguously identified. Nevertheless, two experiments, both involving neutrino interactions, are encouraging. In both, charm may at last have appeared, but even if that is an illusion, the experiments suggest promising lines of research.

One of the experiments was conducted at the Fermi National Accelerator Laboratory in Batavia, Illinois, by a group of physicists headed by David B. Cline of the University of Wisconsin, Alfred K. Mann of the University of Pennsylvania and Carlo Rubbia of Harvard. In examining the interactions of high-energy neutrinos, they found that in several percent of the events, the products included two oppositely charged muons. One of the muons could be created directly from the incident neutrino, but the other is difficult to account for with only the ensemble of known, uncharmed particles. The most likely interpretation is that a heavy particle created in the reaction decays by the weak force to emit the muon. The particle would have a mass of between 2 and 4 GeV, and if it is a hadron, some explanation must be found for its weak decay. Most particles with masses that large decay by the strong force. The presence of a charmed quark in the particle might provide the required explanation.

The second experiment was performed at Brookhaven by a group of investigators under Nicholas P. Samios. They photographed the tracks resulting from the interaction of neutrinos with protons in a bubble chamber. In a sample of several hundred observed collisions one photograph seemed to have no conventional interpretation (see Figure 3).

The final state can be construed as the decay products of a charmed baryon. The process would provide convincing evidence for the existence of charm, if it were not attested to by only one event. A few more observations of the same reaction would settle the matter.

It would be misleading to give the impression that the description of hadrons in terms of quarks of three colors and four flavors has solved all the outstanding problems in the physics of elementary particles. For

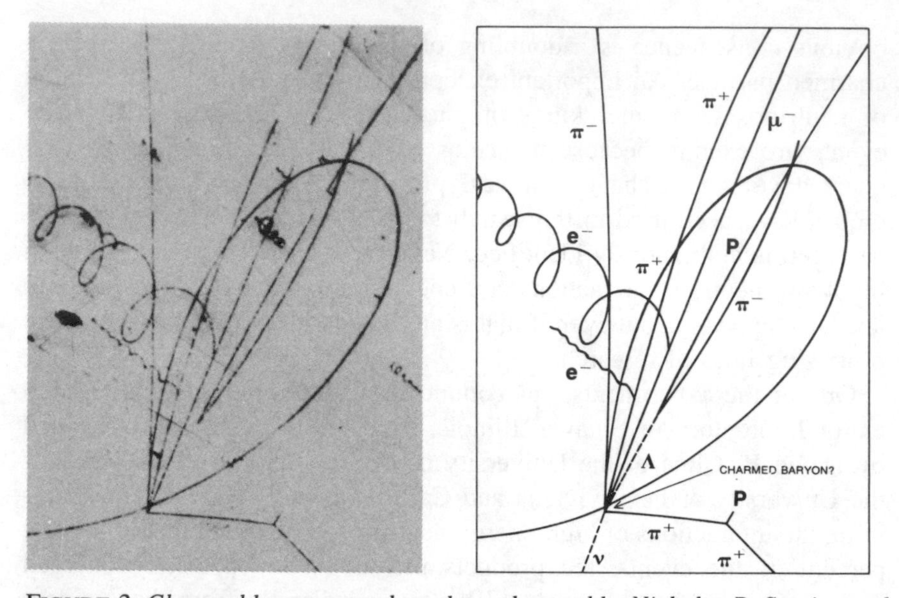

FIGURE 3. *Charmed baryon may have been detected by Nicholas P. Samios and his colleagues at Brookhaven in the aftermath of a collision between a neutrino and a proton. The photograph at left was made in a bubble chamber filled with liquid hydrogen; the particle tracks in the photograph are identified in the diagram at right. The neutrino enters from the bottom left; its track is not visible because only particles with an electric charge ionize hydrogen molecules and leave a trail of bubbles in the chamber. The possible charmed baryon does have an electric charge, but its track cannot be seen either because it is too short; the particle must decay in about 10^{-13} second, so that even at very high speed it does not move far enough to ionize more than a few molecules of hydrogen. The charmed particle decays into a neutral lambda particle, a strange baryon. The lambda particle leaves no track, but its decay products form a vertex that points toward the initial interaction. Four pions and a muon are also created, and two electrons struck by fast-moving particles spiral to the left in the bubble chamber's magnetic field. The presence of the charmed particle is not certain; several other interpretations of the event are possible, and although they are all unlikely, they cannot be excluded.*

example, continuing measurements of the ratio of hadrons to muon pairs produced in electron-positron annihilations have confounded prediction. The ratio discriminates between various quark models, and an argument in support of the color hypothesis was that at energies of from 2 to 3 GeV the ratio is about 2. At higher energy, high enough for charmed hadrons to be created in pairs, the ratio was expected to rise from 2 to about 3.3. The ratio does increase, but it overshoots the mark and appears to stabilize at a value of about 5. Perhaps charmed particles are being formed, but it seems that something else is happening as well: some particle is being made that does not appear in the theory I have described. One of my colleagues at Harvard, Michael Barnett, believes we have not been ambitious enough. He invokes six quark flavors rather than four, so that there are three flavors of charmed quark. It is also possible there are heavier leptons we know nothing about.

Finally, even if a completely consistent and verifiable quark model could be devised, many fundamental questions would remain. One such perplexity is implicit in the quark-lepton symmetry that led to the charm hypothesis. Both the quarks and the leptons, all of them apparently elementary, can be divided into two subgroups. In one group are the u and d quarks and the electron and electron neutrino. These four particles are the only ones needed to construct the world; they are sufficient to build all atoms and molecules, and even to keep the sun and other stars shining. The other subgroup consists of the strange and charmed quarks and the muon and muon neutrino. Some of them are seen occasionally in cosmic rays, but mainly they are made in high-energy particle accelerators. It would appear that nature could have made do with half as many fundamental things. Surely the second group was not created simply for the entertainment or edification of physicists, but what is the purpose of this grand doubling? At this point we have no answer.

Author's note: This article was written just before the discovery of the tau lepton and the revelation that there is a third family of quarks and leptons. Thus, there are at least three times as many fundamental particles in nature as we seem to need to operate spaceship earth. Is there a fourth family awaiting discovery? Its members have already been christened and a series of conferences (in California, where else?) is devoted to discussions of these entirely hypothetical particles. The answer will come from CERN's new electron-positron collider nicknamed LEP. Each family seems to include a light or massless neutrino and antineutrino into which the Z boson must decay. These neutrinos decay modes cannot be detected because neutrinos are too weakly interacting. To count the number of families, all the experimenters need to do is to measure how often the Z boson vanishes without leaving a trace.

The Invention and Discovery of the Charmed Quark

Science often develops in this way: A surprising effect is observed in the laboratory. To explain the new phenomenon, the existing theoretical framework must be enlarged or improved. Thus it was that observed irregularities in the motion of the planet Uranus led to the remarkable and successful prediction of a new planet, Neptune. Galvani noticed that a severed frog's leg twitched when he applied a scalpel to it. This led to an understanding of electric current and to the construction of the first electric battery by Alessandro Volta. The surprising and quite unanticipated discoveries of X rays, radioactivity, and strange particles also fit this pattern of scientific evolution.

There are rare occasions when normal historical order is inverted, when theoretical invention precedes experimental discovery. When Mendeleev formulated his periodic table of the elements, he noticed that it had several blank spaces. He realized that these corresponded to chemical elements which had not yet been discovered, and he worked out their expected chemical and physical properties. A few years later, his predicted elements were found in nature, and were named scandium, gallium and germanium in honor of the countries in which they were found. Mendeleev was recognized to be a great scientist who had the courage of his convictions.

In 1961, Murray Gell-Mann, and Yuval Ne'eman invented a classification scheme which was much like a periodic table of the elementary particles. In this so-called "eightfold way," particles were grouped into simple geometric arrays: hexagons and triangles. Once again, there was a hole in one of these patterns corresponding to a yet undiscovered particle. Not many physicists took this strange new theory seriously. However, in 1964, experimenters in New York's Brookhaven National Laboratory finally discovered Gell-Mann's predicted particle. The discovery of the "omega-minus" forced the heathen to convert and established the eightfold way as scientific dogma.

The success of the eightfold way is now understood in terms of the theory of quarks, just as the success of the periodic table of the elements is understood in terms of the quantum theory of atomic structure. Gell-Mann himself invented the notion of quarks in 1963. (It was done independently by George Zweig, who has since become a neurobiologist.) But, it took a decade for these ideas to become generally accepted.

Gell-Mann postulated the existence of just three kinds of quarks, which were to be the constituents of all the subnuclear particles. Neutrons and protons make up all of the atomic nuclei. These, in turn, are made up of "up quarks" and "down quarks": The proton contains two up quarks and one down quark; the neutron contains one up quark and two down quarks. Short-lived subnuclear particles exist which are made up of three up quarks or of three down quarks.

A third kind of quark is needed to build up the strange particles, and it is called "the strange quark." For example, the "lambda-hyperon" contains one strange quark, one up quark and one down quark; while the "xi-minus hyperon" contains two strange quarks and one down quark. The famous "omega-minus" is made up of three strange quarks.

The basic laws of quark theory say that we can make up a subnuclear particle out of any three quarks. Another species of subnuclear particle, the meson, is made up of one quark and one antiquark. Much of the variety of subnuclear physics comes from the fact that there are three different kinds of quarks to be used.

In 1964, soon after quarks were invented, James Bjorken and I argued that there should exist a fourth kind of quark. We called our conjectured fourth quark "the charmed quark." Not until a decade later was the first particle containing a charmed quark produced and seen in the laboratory.

Our reasoning was once again in terms of a "periodic table": not a table of the elements, nor a table of the subnuclear particles, but a table of quarks and leptons.

Some of the elementary particles are *not* made up of quarks. This is the case for the first elementary particle to be discovered, the electron. It is also true for the various neutrinos. These particles, collectively, are known as leptons. The basic building blocks of matter are believed to be quarks and leptons.

Four kinds of leptons were known of in 1964, forming two pairs closely linked together by the weak force. These were the electron and the electron neutrino, and the muon and the muon neutrino. (The muon is an obese and unstable cousin to the electron weighing about 200 times more.) There were only three known forms of quarks. The down quark had as its partner the up quark, but the strange quark (a weightier relative of the down quark) seemed to be alone and unwed. The analogy between quarks and leptons was incomplete. Just as in Mendeleev's table and Gell-Mann's patterns, there was something conspicuously missing. The up quark surely had to have a rich cousin so that there could be two leptonic couples and two quark couples as well. The missing quark was our invented "charmed" quark. (More recently the party has grown, for there are now three married couples of each, although the sixth quark continues to elude the observations.)

By 1970, there was still no shred of experimental evidence for the existence of the charmed quark. Nonetheless, with my colleagues John Iliopoulos and Luciano Maiani, I returned to this fascinating speculation. We produced convincing, but very indirect, evidence that the charmed quark had to exist. The theory without charm was very unsymmetrical, predicting certain kinds of asymmetries in nature which were known not to be there. The charmed quark would restore the symmetry, thereby removing disagreement with experiment.

By April 1974, I became very much annoyed by the failure of experimenters to find the charmed quark, which I knew must exist. At a meson spectroscopy meeting in Boston, I predicted that charm was about to be discovered, and I promised to eat my hat if it were not found before the next meeting of that series. In November of that year, the discovery of a new particle was announced simultaneously by workers at Brookhaven and at a machine called SPEAR in California. SPEAR is a larger version of the electron-positron colliding beams which were pioneered at Frascati. Indeed, scientists at Frascati were able to confirm the existence of the new particle soon after it was found.

The East coast group called the new particle J while the West coast group called it psi. Today it bears the double name J psi. The leaders of the two groups, Burton Richter and Samuel C.C. Ting, shared the 1976 Nobel Prize in physics for their discovery of this important new particle.

Soon after the discovery of the J psi, there were many arguments about what this new particle was. My colleagues and I believed that it was a meson made up of a charmed quark and its antiquark. But, if this were true, there should exist other new particles containing just one charmed quark. Where were they?

Nick Samios, at Brookhaven, was the leader of the group that found the first omega-minus in 1964. In early 1975, his group reported the first observation of a charmed particle, a particle consisting of an up quark, a down quark, and a charmed quark. It was produced, in a large bubble chamber, by a beam of high-energy neutrinos. However, he could identify only one example of the production of a charmed particle. This was not enough to convince the physics community of the existence of charm.

In the spring of 1976, a group of physicists working at SPEAR published a paper reporting that they could not find any sign of charmed particles. Now, I became really angry, and at a meeting in Wisconsin I insisted that the experimenters go back to their laboratory and find the new particles that simply had to be there. My arguments were effective. Only two weeks after the meeting, my friend Gerson Goldhaber telephoned from California to tell me that his group at SPEAR had finally discovered the elusive charmed mesons. Soon, they were shown to have precisely the properties that we had predicted in advance.

It was a long wait, from 1964 when the idea of charm was born, to 1976 when it became an established fact. What joy it was for me to participate in such an adventure! Needless to say, when the conference on meson spectroscopy met again in 1976, I was not obliged to eat my hat. To the contrary, the organizers of the conference distributed candy hats to all of the participants. It was they who ate their hats.

Antineutrinos and Geology

The exotic particles of high-energy physics can sometimes prove to be very important in distant disciplines. Pions and muons were identified as cosmic ray components only 35 years ago. Today, beams of pions are routinely employed in the treatment of certain kinds of cancer, and cosmic-ray muons have been used to ''X-ray'' the Egyptian pyramids in the search for hidden treasures. Neutrinos, the ghostlike particles invented by Wolfgang Pauli in 1930, were first observed in the laboratory in 1955. They seemed to have no practical significance, but things have changed.

The discipline of neutrino astronomy is becoming more and more important. Neutrinos from the sun have been detected and measured by an experiment located in a deep gold mine in South Dakota. The results are *not* in agreement with theoretical expectations. There are things about the sun that still mystify us. Cosmic rays, when they impinge on nuclei of the atmosphere, are another source of neutrinos, and these have been seen too. They produce an undesirable and confusing background in the yet inconclusive search for proton decay. Another astronomical source of neutrinos is from those titantic stellar explosions known as supernovae. The neutrino burst that accompanies a supernova lasts for only a few minutes, and it can be detected if the supernova takes place near enough to us. Unfortunately, such events take place only about once a century. The search for supernova neutrino bursts requires patience beyond the call of duty.

While the sun is a powerful source of *neutrinos,* our earth is a weaker but nearer, source of *antineutrinos*. Loosely speaking, neutrinos are produced by any process that assembles large nuclei from smaller ones: like the fusion of four hydrogen nuclei to form helium—the process that fuels the sun. Antineutrinos, on the other hand, are produced by any process that breaks larger nuclei into smaller ones. Nuclear power plants and nuclear weapons are plentiful sources of antineutrinos since they make use of the fission of large nuclei. However, antineutrinos are continually being manufactured by natural processes as well; by the radioactivity of certain elements which are present in the earth's crust: potassium, rubidium, thorium, and uranium. The fact that naturally radioactive materials exist upon earth has had a remarkable impact on human civilization.

The first hint of the existence of what are now known as weak interactions was the discovery of radioactivity in the late nineteenth century. Soon thereafter, Madame Curie discovered the radioactive children of natural uranium: radium and polonium, and Rutherford discovered that radioactivity is a form of transmutation. The long-sought dream of the alchemists turned out to be a naturally occurring phenomenon. Using the products of radioactivity as a microprobe of the atom, Rutherford in 1911 demonstrated the existence of the atomic nucleus. The study of natural radioactivity in the early decades of this century has led us to our present deep knowledge of atomic structure.

Moreover, the occurrence of radioactive materials on earth has given us the ability to estimate the age of our planet. Radioactive potassium, for example, has a half-life of about two billion years. It decays into a common form of calcium. From the known concentrations of potassium and calcium on earth today, and from similar considerations of other radioactive elements, we have deduced that the earth is about four and a half billion years old.

The process of radioactivity produces antineutrinos, but in addition, a considerable amount of heat is released. The sum total of this radiogenic heat produced by all of the radioactive elements in the earth's crust is estimated to be 20 million-million watts (or 20 terawatts). This is to be compared to the five terawatts released by the burning of fuel by all the people and factories of the world, and it is 100 times larger than the total American electric power production. This source of heat plays a significant role in the history of earth. Were our planet free of radioactive materials, the earth's core would have congealed eons ago and the geomagnetic field would have been extinguished. There would have been no such thing as a magnetic compass, and the fifteenth-century explosion of maritime discovery might not have taken place. Columbus discovered America in 1492 because of the existence of natural radioactivity! Even more important than this is the fact that the earth's magnetic field is its shield against the devastating effects of cosmic rays: a nonradioactive, and hence nonmagnetic, earth could never have given birth to life as we know it.

In a publication written with Lawrence Krauss and David Schramm, we compute the number of antineutrinos produced by the earth's known store of radioactive minerals. Our estimate is likely to be low since we really do not know how much radioactive material lies inaccessibly deep within the earth. We find that some 10 million antineutrinos emerge from each square centimeter of the ground in each second.

With considerable difficulty, it is possible for a sensitive experiment to detect and measure these antineutrinos. However, the experiment must be done deep underground so that it is protected from cosmic rays. Perhaps the experiment will be done in the new underground laboratory beneath the Gran Sasso.

It is an important experiment for precisely the reason that we cannot calculate the antineutrino flux with any precision. By measuring antineutrinos, we may unravel two of the deepest (please excuse the pun) mysteries of geological science: what is the chemical composition of the bulk of the earth, and how much heat is being produced by its radioactive constituents? It would demonstrate once again that high-energy physics is not only interesting for its own sake (as it most certainly is), but it may be used as a powerful and effective tool for all of the other sciences: in this case, for the study of the last earthly frontier—the center of the world.

Author's Note: In 1987, there was a supernova in the Large Magellanic Cloud. Its neutrinos were observed at laboratories in Japan and the United States. It was not so much a result of patience as serendipity. Scientists searching for proton decay saw a supernova instead.

Elementary-Particle Physics as a Waste of Time and Money

Elementary-particle physics has become recondite, expensive, and totally impractical. Kaons, muons, neutrinos and their ilk will probably never have any impact on future technologies. The study of classical mechanics is essential to the building of weather satellites. Solid-state physics brought us the transistor. Nuclear physics and the relativistic formula, $E = Mc^2$, led to atomic energy. Quantum mechanics is at the root of modern chemistry and biology. But, particle physics seems to be an endless road to nowhere. Each year marks the discovery of more and more particles with absurdly short lifetimes. Alphabets are exhausted in their taxonomy. As we get deeper into the subnuclear world, things seem to get even more complex, irrelevant, and even random. Many of the currently fashionable "ultimate building blocks of nature"—like muons and charmed quarks—do not even seem necessary to build up the universe we live in. They seem to be put in for spite. A knowledge of elementary-particle physics is not necessary nor even useful to most practicing physicists. And, many highly qualified particle physicists cannot even find jobs in their chosen specialty. If, according to Ruderman, cosmology is the saddest of sciences, then surely elementary-particle physics is the most futile.

Puzzle solving and monument building are characteristically human activities. Particle physics has much in common with each. Our discipline is often compared to a jigsaw puzzle, wherein some pieces have been neatly fitted together and the big picture is just being glimpsed. But there remain many large holes, and pieces that simply don't seem to fit anywhere. Who can resist the challenge of a puzzle whose pieces are the *very basic stuff* of nature, one which has been worked on for thousands of years? How personally satisfying it could be to put one more piece in its place. The puzzle is itself a monument. Like the pyramids, the music of Bach, or the conquest of Everest, the accomplishments of particle physics remain a permanent part of our legacy to the future. We may recklessly squander our fossil wealth, but our knowledge of charmed hadrons is eternal.

Builders of pyramids and of cathedrals had considerable political clout. Their societies delivered a significant portion of their wealth to the monument builders, thereby neglecting many activities that we might consider to be far more important. Our society is simply not so committed to monument building, and even less so to puzzle solving. How do we explain the construction of accelerators costing hundreds of millions of dollars? Is it just friendly competition with the Russians or the Europeans? Is it the pure spirit of scientific endeavor, without hope of tangible gain? Is it the political pressure of particle physicists and construction unions?

Although remote, there is the possibility that particle physicists will stumble onto something "useful." For a brief time, the false hope was held that muons could be the key to fusion devices. It is rumored that Russians hope to modify particle accelerators for military purposes, and *The New York Times* has prophesied ships powered (somehow) by magnetic monopoles. Nonetheless, I believe that our discipline *per se* will turn out to be utterly "useless."

On the other hand, particle physics can be immensely important to society in an indirect way, by the phenomenon called spin-off. Experiments often involve new challenges in electronics, computer science, cryogenics, or other fields. Many advances in these disciplines were initiated within high-energy physics laboratories. Accelerators have been found to have applications outside of their original domains: nuclear physics, clinical medicine, and fusion research.

In addition to technological spin-off, there is methodological spin-off. Relativistic quantum mechanics, the language of particle theory, has much in common with the theory of collective phenomena (many-body physics) and of nuclear theory. Advances in one field generally signify and stimulate advances in the others. Ultimately, such cross-fertilization spreads across all of the often artificial boundaries between sciences, to the enrichment of all.

People are also spun off. It can be argued that particle physicists attack some of the most obstinate and perverse problems known. Experiments that were once technologically impossible are done and then become routine. Calculations that once took a year can now be done in minutes. Not only do particle physicists ask and answer difficult questions, but they are also faced with far more challenging questions of what are the right questions to ask. Consequently, it is not surprising that particle physicists who switch fields often succeed brilliantly: examples in biology, medicine, computer science, mathematics, and political science come to mind. Elementary-particle physics builds character and keen minds.

Particle physics is the current name of the search for the ultimate constituents of matter and for the rules by which they combine. Particle physics is a new science, but the search can be traced to antiquity. Several times, scientists had thought they had identified their prey, only to discover that there were underlying simpler structures. The format is always the same. An entity is proposed to be elementary; many varieties of the entity are observed; a suggestive systematics is revealed; and finally, the entity is recognized to be a composite of more elementary parts and a dynamics is devised which explains how the parts are held together.

The existence of atoms was proven to many by the work of Dalton in the early nineteenth century. Elements combined to form compounds in precisely fixed ratios. The only simple explanation of the "law of definite proportions" involved the existence of minimal units of each element—atoms. Other arguments for atoms followed, involving the kinetics of gases or the phenomenon of Brownian motion.

There were as many different kinds of atoms as there were elements known. The diverse atoms had widely different masses and properties. It could not have seemed less promising that a system of order would emerge. Mendeleeff was nonetheless convinced that nature had a simple secret to tell; he believed that it is "the glory of God to conceal a thing [but] the honor of kings to search it out." Six years passed between the formulation of his periodic table, and its experimental vindication. The discovery of the elements gallium, germanium, and scandium, with just the properties that were predicted, silenced those who felt that the descriptive successes of the periodic table were mere coincidences.

In the early twentieth century, Rutherford demonstrated that the atom was not an elementary particle. By observing the elastic scattering of alpha particles off gold, he concluded that most of the atomic mass resided on its small dense nucleus. Each atom consisted of a pointlike constituent ''parton'' surrounded by a cloud of electrons. The dynamics of the system was to be explained with the development of quantum mechanics, and the mysterious successes of Mendeleeff's table were completely understood.

Is it then the atomic nuclei that are elementary? Again, this was not to be. There were too many different nuclei for one to believe they were all elementary—at least one nucleus must correspond to each element. The discovery of isotopes showed that the number of district nuclei was indeed larger than the number of elements. The mass of each nucleus was seen to be an approximately integer multiple of the mass of the hydrogen nucleus. Moreover, Mosely showed that the chemical properties of each element could be put into correspondence with another integer, the atomic number. The observed pattern of atomic weights and atomic numbers was a convincing clue that the nucleus was a composite system. However, attempts to build nuclei out of the simplest nucleus (the proton) and of electrons simply failed.

The single experimental discovery most relevant to the elucidation of nuclear structure was that of the neutron. It was suddenly obvious (to the open-minded) that nuclei were built up of two kinds of ''nucleons'': protons and neutrons. The number of constituent nucleons is the atomic weight; the number of protons is the atomic number. Simple as pie. The final stage of this format is almost complete. We do have a modestly successful theory of nuclear structure: the so-called strong interactions are responsible for nuclear binding, and many detailed features of nuclear physics are explicable. For a time it was even believed that the basic building blocks of matter had at last been found.

George Gamow wrote in 1947:

> ''But is this the end?'' you may ask. ''What right do we have to assume that nucleons, electrons, and neutrinos are really elementary and cannot be subdivided into still smaller constituent parts? Wasn't it assumed only half a century ago that the atoms were indivisible? Yet what a complicated picture they present today!'' The answer is that, although there is, of

> course, no way to predict the future development of the science of matter, we have now much sounder reasons for believing that our elementary particles are actually the basic units and cannot be subdivided further. Whereas allegedly indivisible atoms were known to show a great variety of rather complicated chemical, optical, and other properties, the properties of elementary particles of modern physics are extremely simple; in fact they can be compared in their simplicity to the properties of geometrical points. Also, instead of a rather large number of "indivisible atoms" of classical physics, we are now left with only three essentially different entities: nucleons, electrons, and neutrinos. And, in spite of the greatest desire and effort to reduce everything to its simplest form, one cannot possibly reduce something to nothing. Thus, it seems that we have actually hit the bottom in our search for the basic elements from which matter is formed.

The picture was soon to change. Yukawa prophesized that there must be particles to mediate nuclear forces just as photons mediate electromagnetism. His "mesotrons" were first observed in 1948, and were eventually to be called "pions." It became clear that nucleons had none of the simplicity of geometrical points. Like atoms and nuclei, they have a wide variety of properties. Nucleons bear "anomalous" values of magnetic moment, not those predicted by Dirac's theory of pointlike particles. They have a spread-out distribution of electromagnetic properties, such as one would expect of particles with a definite size. And, there are many, many particles like nucleons and pions that partake in the nuclear force. Strange particles were first detected in 1947.

Then, in 1954, an excited state of the nucleon was discovered: the first "resonance," with a lifetime far too short to be directly measured. The work "hadron" was coined to signify any of these strongly interacting particles. By the early 1960s there were more than 100 observed hadrons. All these particles were clearly related to neutrons and protons. All seemed equally elementary or not elementary. This population explosion, and the nuclear democracy it led to, was the first clear hint that there was another layer to the onion.

The notion of isotopic spin offers a partial classification of the hadrons. Its extension to the approximate symmetry scheme known as "the eightfold way" played somewhat the same role for hadrons that the periodic table did for atoms. It too was dramatically predictive. The observation of the Ω^- in 1964 with just the expected properties ended all doubt. Quarks were invented in 1963 by Gell-Mann and Zweig to provide a small set of fundamental entities (three), out of which the many hadrons could be built. The quark theory gave a natural explanation for the observed systematics of hadrons. Departures from exact symmetry merely reflected the mass spectrum of the three quarks.

Moreover, an updated version of the Rutherford experiment provided more direct evidence for the existence of quarks within nucleons. The energy of the bombarding particles was higher, and they were electrons not alpha particles, but the interpretation was much the same. Each nucleon seemed to consist of three pointlike "partons" somehow held together.

Today, it is generally believed that the partons are quarks. A non-Abelian gauge theory—chromodynamics—is responsible for the force between quarks. More indirectly, chromodynamics is at the root of the strong interactions between hadrons. The situation has a close parallel in the atom, where it is naked electrodynamics that bind the electrons, but the much less direct van der Waals force which describes the forces between atoms. Strong interactions are now regarded as a secondary consequence of the far more fundamental chromodynamic force.

Thus, we now use quarks and leptons as the most fundamental building blocks of nature. Whether or not we have descended to the ultimate stratum of the microworld is entirely unclear. Certainly, the leptons behave in all ways as if they were truly elementary. Quarks are more puzzling. According to the current dogma, they are not isolable. How can we demonstrate further structure in the quark if we cannot even separate it from its hadron? Perhaps the impossibility of finding quarks is nature's way of letting us know we have reached the end of the line. But Sir Denys Wilkinson responds:

> But we know that we could not accept that. We know that whether or not we can get at quarks we shall continue to talk about them and discuss their anatomy and domestic economy. We shall talk about their form and structure and even though perhaps we can never know, our bowels will tell us that this is the way it has to be.

The first symptom of nonelementarity—for atoms, for nuclei, and for hadrons—was always the same: an uncontrolled growth of the number of distinct entities alleged to be elementary. As we shall see, there is evidence that it has happened again. Let us approach the current situation in the historical context of lepton-hadron symmetry. Until 1932, there was just one hadron (the proton) and one lepton (the electron). A new age was ushered in by the almost simultaneous discovery of the neutron and invention of the neutrino. Actually, the neutrino was not directly detected until 1956, but the arguments for its existence in the early 1930s were iron-clad. Now that we know that nucleons are made up of two kinds of quarks, we can refer to the period 1932–1944 as the age of two quarks and two leptons.

Once again, two dramatic discoveries marked the beginning of a new age. The muon was first seen in 1938, but was mistaken for a hadron. The situation was clearing up in the mid-1940s when strange particles were also first seen. Electrons, neutrinos, muons, up, down, and strange quarks: these are the fundamental particles in the third age of lepton-quark symmetry 1944–1962.

In 1962, the muon's own neutrino was found to be different from that of the electron. This led to the only significant period of lepton-quark asymmetry: the interregnum of 1962–1974. During this period, physicists became so accustomed to the idea of just three quarks, that arguments for the existence of a fourth were generally ignored. This age ended in November 1974 with the simultaneous discovery of the J/ψ at Brookhaven and at SPEAR. It was soon to be realized that this particle is a combination of a charmed quark and its antiquark: charmonium, after the word positronium describing an analogous leptonic configuration.

The fourth age of lepton-quark symmetry was remarkable in that it marked a kind of parity between relevant particles and irrelevant ones. Two quarks and two leptons suffice to build the universe we live in. There seems to be no need for charmed and strange quarks, nor for muons and their neutrinos. Each of the relevant particles plays an important role in our day-to-day world. Even the neutrino is essential: it was necessary for the synthesis of the elements and it remains necessary to power the sun and the stars. For a time, it seemed as if Nature simply used exactly twice as many building blocks as were needed, and this grand doubling could itself be regarded as a key to the puzzle.

Unfortunately for such a notion, we have passed into the fifth age of lepton-quark symmetry. The most conservative explanation of Lederman's recent data is the existence of a fifth quark some three times heavier than the charmed quark: unpredicted, and perhaps unwanted, but it is there. SPEAR and DORIS have presented incontrovertible evidence for the existence of a fifth lepton, some 17 times heavier than the muon. The fourth age of lepton-quark symmetry was very short-lived.

Are we done with just five building blocks of each kind? Theoretical arguments strongly suggest not. The new lepton, called the taon, probably must have its own neutrino. It almost certainly cannot make do by sharing one of the other neutrinos. Similarly, the existence of an unpaired quark could easily lead to trouble in the hadronic domain. The absence of flavor-changing neutral currents, which is now more-or-less established, generally requires the quarks to come in pairs. We are almost certainly not yet done. The idea of lepton-quark symmetry, retrospectively so useful, may or may not continue to hold true. At present, it is no more than a clue. What is needed is a theory which tells us how many quarks and leptons there must be.

One thing seems clear: there are again simply too many fundamental building blocks: at least a dozen and probably more. One is sorely tempted to speculate that there are more layers to the onion, and that quarks and leptons will someday be found to consist of simpler and fewer things. Such a notion has been championed by many Chinese physicists. I would propose that these purely hypothetical building blocks of all matter be called ''maons,'' to honor the late Chairman Mao who insisted upon the fundamental unity of nature.

Maons are for the future. Meanwhile, we have many varieties of quarks and leptons of which only four currently seem to be relevant. The useful ones are those, which by accident of birth, are the lightest. Not only are we descended from slime and resident upon an inconspicuous and unimportant sliver of the universe, but the universe itself is built up out of a small minority of the most fundamental entities: the runts.

Does Elementary-Particle Physics Have a Future?

Physicists deal with an incredible range of distances, from the inconceivably small Planck scale of 10^{-33} cm to the incomprehensible size of the visible universe, 61 powers of ten larger. Arranged sequentially upon this cosmic ruler are the many disciplines of science: particle, nuclear, and atomic physics, then chemistry, biology and geology, and finally astronomy and cosmology. All these fields are ultimately quantum mechanical, and quantum mechanics began with Niels Bohr.

Typically, Bohr began not at the beginning of the ruler nor at the end, but at the muddle in the middle, the atom. He saw that the classical rules could not describe Rutherford's nuclear atom so he invented new ones which did. His rules evolved into a theory which explained the mysteries of the atom and the successes of the periodic theory. But, quantum mechanics is a greedy master which admits of no competition. Quantum rules must rule the whole ruler.

The quantum atom led to the quantum nucleus, which is built up of nucleons held quantum-mechanically together. Nucleons themselves are built up of quarks, whose quantum nature is demonstrated by the discovery of dozens of "stationary states" of the proton. The smaller things are, the more they are quantum mechanical. For this reason, it took physicists decades to recognize these states for what they are.

For objects the size of a mountain and larger, gravity is the dominant force and the relevance of quantum mechanics seems to fade. But, stars shine because of a complex interplay of all of the forces of nature in an essentially quantum-mechanical game. Quantum mechanically as well, the hot early universe of 10 billion years ago cooked up many of the light nuclei now about us. Things work out so well that we know that the laws of quantum mechanics were the same then as they are now. Truly, *plus ça change, plus c'est la même chose.*

In the earliest moments of the formation of our universe, things were so very hot that particle physics and its bestiary of quarks and leptons reigned supreme throughout the universe. The ruler has curled up upon itself—it is no ruler at all but a snake swallowing its own tail, Ouroboros. The physics of the microworld is the physics of the entire cosmos. The large and the small are one. Our earthly accelerators are at once microscopes of supernal resolution, and miniature replicas of the greatest accelerator of them all, the entire universe.

The unification of the small and the large is twofold. Gravity itself must be a quantum-mechanical theory: Bohr and Einstein must ultimately be reconciled. Quantum gravity is dominant only at times so early that the temperature of the universe approaches the Planck mass, or conversely, at the very smallest distance scale of the Planck Compton wavelength. Once again the snake swallows its tail, but at energies and at distances far removed from any conceivable laboratory but the universe itself.

Bohr's modest domain of atomic sizes has been expanded and expanded by the arrogant reductionism of today's physical scientist to cover the whole shebang. Mysteries still confront us at the largest distance scales: What is the dark matter of the universe? Why do globular clusters seem to be older than the universe itself? How did the galaxies form? And so on. But, we have made remarkable progress at the smaller scales. We have what appears to be a correct, complete and consistent theory which describes all the known phenomena of the microworld: quantum chromodynamics and the electroweak theory.

On vacation upon the lovely isle of Jamaica, I discovered that Jamaicans know but one variety of cheese. It is used on pizzas, in sandwiches, and in omelets. It is called standard cheese. So it is in particle physics. We have only one theory that works, and a very good theory it is. Quantum chromodynamics and the electroweak theory comprise our standard model of particle physics. It is used in cosmology, astrophysics, nuclear physics, and in particle physics. There is really no choice.

No known phenomena suggest structure beyond the standard model. No measured quantity contradicts the standard model. There are no internal contradictions, and there are no loose ends. Yet, the standard model appears to no one to be a satisfying conclusion to the search of the particle physicist. For one thing, quantum chromodynamics (QCD) is not yet as predictive as one might hope. It has not yielded the observed mass spectrum of the hadrons. Presumably, this is a computational question which will be resolved by further study and future development of computer systems. There are more fundamental puzzles outstanding.

Why is the gauge group what it is? Why are there three families of fundamental fermions? Is there a Higgs boson, or what? Aren't 17 basic particles and 17 arbitrarily tunable parameters far too many? How about a quantum theory of gravity?

Have you noticed that we have made the great leap sidewise? Once upon a time, we particle physicists could honestly claim to be studying the ultimate structure of ordinary matter, from atoms, to their nuclei, to the garden varieties of quarks. Things began to change about 40 years ago with the discovery of the muon. I.I. Rabi, like other New York physicists fond of Chinese food, is said to have said upon hearing about the muon, "Who ordered that?" It is among the very few questions that Rabi's student, Julian Schwinger, hasn't answered. Schwinger's own student, yours truly, doesn't yet know.

Today the muon has been joined by the tau lepton, strangeness, charm, top and bottom. We no longer study the structure of the atom, for two-thirds of our particles have nothing whatever to do with mundane matter. They are exotica found only at large accelerators or in the debris of cosmic-ray collisions. They have no more to do with "ordinary physics" than do elements numbers 108 and 109 have to do with "ordinary chemistry." We seem to be following an endlessly difficult and expensive side issue. Have we lost the thread of relevance?

We have not at all lost our way. It is just that we are not yet there. The standard theory cannot be our final answer just because it cannot justify the great leap sidewise. Like my Harvard predecessor, Percy Bridgeman, I (and, surely, we) have a quite unjustifiable faith in ultimate simplicity, in the existence of a one and true theory: unjustifiable but always justified by the remarkable progress in our discipline. Today's side issue will one day be central. The muon will find its essential place in the sun, along with its curious brethren.

Yes, my children, elementary-particle physics does have a future. Yet, today it is threatened, and its exposure may be greater than ever before.

Our discipline is threatened by the recent divorce between particle experiment and particle theory. Perhaps it all began with quantum chromodynamics, an apparently correct theory underlying the quark structure of nucleons and the nuclear force itself. It is not merely *a* theory, but within a certain reasonable context, it is *the* unique theory. In principle, QCD offers a complete description and explanation of nuclear physics and of particle physics at accessible energies. While most questions are computationally impossible to answer fully, the theory has had very many qualitative (and, a few quantitative) confirmations. It is almost certainly "correct." QCD is not the threat I have in mind. It has *not* produced a divorce between experiment and theory—indeed, it has led to closer coordination and cooperation between experimenters and theorists. Yet, it has planted a seed that has blossomed elsewhere. It suggests and affirms the belief that elegance and uniqueness can be criteria for truth. I believe in these criteria. But, they must be reinforced by experiment. I agree with Lord Kelvin when he says,

> When you can measure what you are speaking about and express it in numbers, you know something about it. And when you cannot express it in numbers, your knowledge is of a meagre and unsatisfactory kind. It may be the beginning of knowledge, but you have scarcely in your thought advanced to the stage of a science.

By this criterion, too, QCD is a science. But, can the same be said of the superstring and its ilk?

Quantum mechanics is contagious, and gravity must be framed within its context. Some of my theorist friends feel that they have come upon the unique quantum theory of gravity: a supersymmetric system of strings formulated within a 10-dimensional space-time. Most of the physics of the superstring lies at forever inaccessible energies, up around the Planck mass. Within its context, the theory is unique. It may even be finite and self-consistent. It seems capable of describing the low-energy phenomena which we can observe in the laboratory, but it is hard to prove that it really does. In principle, it predicts what particles exist. In principle, the number of tunable parameters is reduced to zero. In practice, however, it has made no verifiable prediction at all, and it may not do so for decades to come. The string theorist has turned towards an inner harmony. But, can it be argued that elegance, uniqueness, and beauty define truth? Has mathematics supplanted and transcended experiment which has become irrelevant? Will the mundane problems which I call physics, but which they call phenomenology, simply come out in the wash in some distant tomorrow? Is further experimental endeavor not only difficult and expensive, but *unnecessary* and *irrelevant*? Perhaps I have overstated the case made by string theorists in defense of their new version of medieval theology where angels are replaced by Calabi–Yau manifolds. The threat, however, is clear. For the first time ever, it is possible to see how our noble search could come to an end, and how Faith could replace Science once more. Personally, I am optimistic. String theory may well dominate the next 50 years of fundamental theory, but only in the sense that Kaluza–Klein theory has dominated the past 50 years. Perhaps we should turn our attention to the past for guidance about the future.

The search for the ultimate constituents of matter has had a cyclic history passing from chaos, to order, to a new level of structure, and to chaos once more. We have passed through four such cycles in recent history: atoms, their nuclei, hadrons, and now quarks and leptons.

Recall the search for new chemical elements, which precedes Lavoisier and continues to this day. Figure 1 shows the growth of the number of known species with time. The most recent additions, numbers 108 and 109, were produced and observed in Germany in the 1980s. The curve increases almost linearly over two-and-a-half centuries, showing spurts

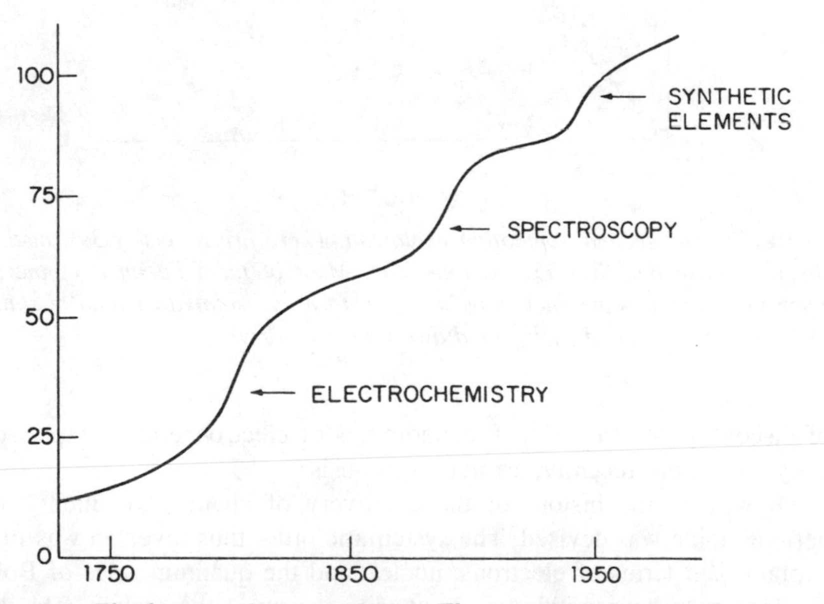

FIGURE 1. *The first population explosion. The growth of the number of known chemical elements over the past three centuries. The curve is roughly linear, with superimposed spurts of discovery resulting from technological developments. Elements 108 and 109 were produced in the 1980s. Will there be more?*

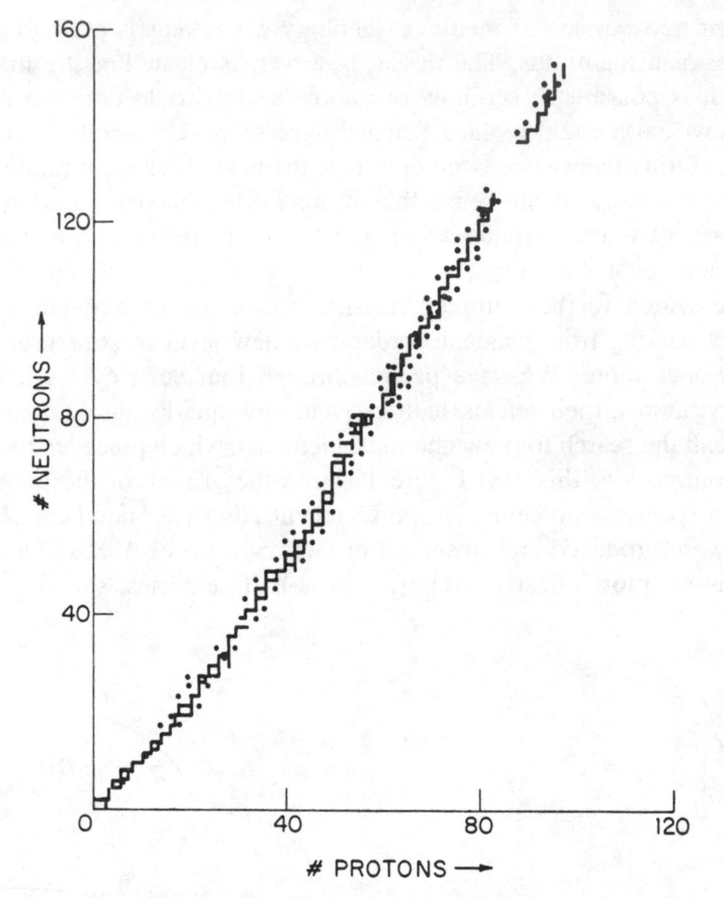

FIGURE 2. *The second population explosion. There are precisely 399 nuclear species whose half-lives exceed one year. Most of them lie on a connected "peninsula," but some, including uranium, form an isolated "island of stability." Is there a second such island awaiting discovery?*

of discovery due to the new technologies of electrochemistry, spectroscopy, and more recently, artificial synthesis.

Midway in the history of the discovery of chemical elements, the periodic table was devised. The systematic order thus revealed was fully explained in terms of electrons, nuclei, and the quantum rules of Bohr. Nuclei were soon identified as composite structures themselves. The discovery of isotopes showed that there are far more nuclear species than there are chemical elements. Indeed, Figure 2 shows a plot of the 399 nuclear species whose half-life exceeds one year. (My colleague, Roy

Glauber, had guessed 400.) Clearly, there are far too many nuclides for them to be elementary. The integral values of Z discovered by Moseley, and the almost integral values of A, represent the second level of order. Structure awaited the discovery of the neutron in the year of my birth.

Things were beginning to look simple. In the late 1940s, Gamow wrote that the elementary particles were only three in number: nucleons, electrons, and neutrinos. These were the pointlike particles that were the basic building blocks of all matter. Alas, it was not to be so simple. Pions and strange particles were discovered. As large accelerators were deployed, there was a virtual population explosion among the nuclear particles, or hadrons. Figure 3 shows the time evolution of the number of known hadron types. The curve rises approximately linearly from a few to more than 100 in a time interval of only several decades. The curve is reminiscent of the explosion of known atomic species, but compressed in time by a factor of 10. As before, wiggles in the curve

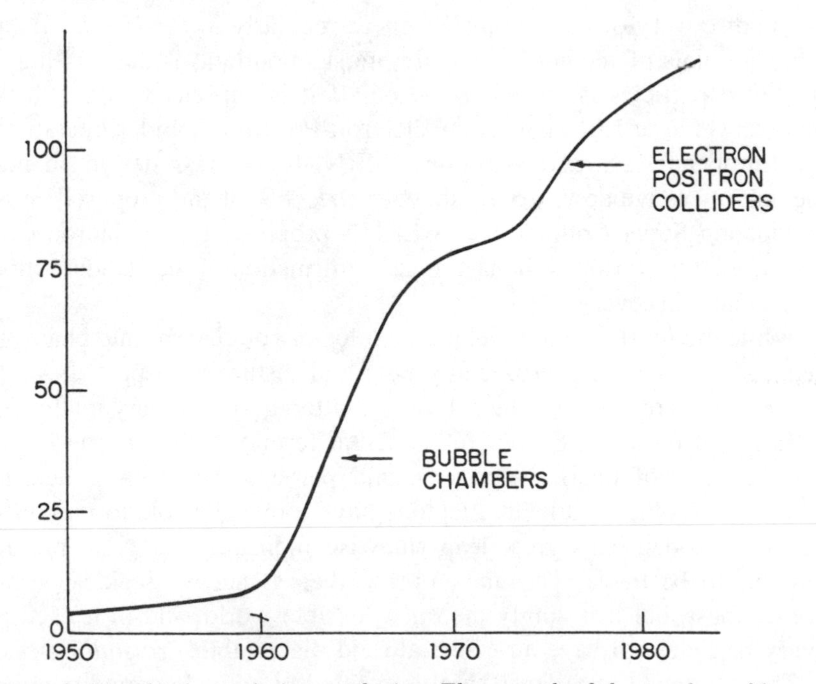

FIGURE 3. *The third population explosion. The growth of the number of known hadron multiplets over the past three decades. Again, the curve is roughly linear with spurts due to the deployment of new instruments. There are now about as many hadrons as there are chemical elements. Soon there will be more.*

correspond to developments in experimental technology: bubble chambers and electron-positron colliders.

Once again, midway in the evolution of the curve, a new level of order is discovered: the eightfold way of Gell-Mann and Ne′eman. New particles with specific properties were predicted by the theory. The discovery of the famous Ω^- particle played the same role in this third cycle that the finding of scandium, germanium and gallium did for the first. Gell-Mann was no more mad than Mendeleev. For a third time, the appearance of systematic order led to the revelation of a new layer of structure. The hadrons were found to have all the attributes of composite systems: they were shown to be made up of quarks.

Today's candidate elementary particles have endured a quieter population explosion than their predecessors. The time evolution of the number of known quark species, leptons, and force particles is shown in Figure 4. There are, in our standard theory, just 17 building blocks, and 14 have been unambiguously discovered in the laboratory. Indirect evidence for the existence of the tau neutrino is compelling. Muted reports of the discovery of the top quark appear regularly in *Physics Today* and other journals of record. By far, the most important of the missing but predicted particles is the Higgs boson. If it is sufficiently light, it will be discovered at LEP, the Large Electron-Positron Collider operated by the European research consortium CERN. If its mass lies in an intermediate mass window, it will show in the U.S. at the proposed Superconducting Super Collider or at CERN's proposed Large Hadron Collider. The Higgs boson is the last great confirmation of the standard model that awaits discovery.

While the most recent population explosion, or descent into chaos, has been gentle, it is of a profoundly new and disturbing aspect. In earlier cycles, we were studying the nature of matter, quite ordinary matter such as is found on earth. Of our fundamental fermions, this is true for just two varieties of quarks, electrons, and perhaps the electron neutrino. None of the other quarks and leptons have a relevant role to play in the standard model. This great leap sidewise indicates that great progress remains to be made. The one-and-true theory that we seek is perhaps not the best, but it is surely the *only* possible world, and in it each and every particle will have an essential (and discernible!) *raison d'etre*.

The next level of order is surely revealed by today's periodic table of quarks and leptons, Figure 5. The fundamental fermions form families, each with the same recurrent pattern of strong and electroweak proper-

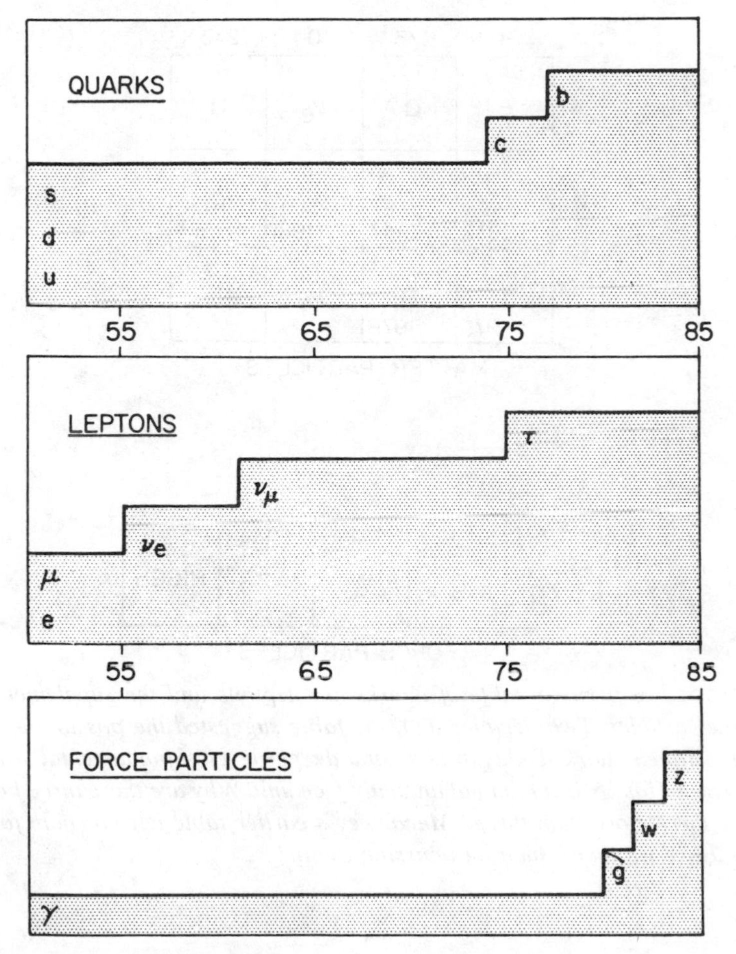

FIGURE 4. *Today's population explosion of fundamental particles. According to our canonical theory, only one particle of each type remains to be found. The top quark and the tau neutrino are relatively easy. The search for the Higgs boson is the outstanding challenge to defenders of the standard model. Will it be found? Will something unexpected be found?*

ties. The pattern is explained in terms of grand unification, but the reason for the seemingly superfluous replication of families remains obscure.

From the time of Bohr to the present, particle physics has progressed enormously. Many, indeed most, of the problems of yesteryear have been solved. To a large extent, our success is the result of experimental discovery. Generally, the essential empirical basis to our theory is the result

-1	-1/3	0	2/3
e	d	ν_e	u
μ 1938	s 1947	ν_μ 1961	c 1974
τ 1975	b 1976	ν_τ	t

MATTER PARTICLES

γ	g 1980	W 1983	Z 1984	HIGGS

FORCE PARTICLES

FIGURE 5. *The periodic table of quarks and leptons and the supplemental list of force particles. Two decades ago, the table suggested the possible existence of the charmed quark. Today, it hints at a deeper level of fundamental structure. Each row or family is a minimal anomaly-free unit. Why are there three families, or, are there more than three? Mendeleev's earlier table left no room for inert gases. Have we made such an omission again?*

of patient, even plodding, endeavor. It is the accumulated knowledge due to very many scientists, all too few of whom will be remembered for a particularly startling or significant discovery. It has been a history, though, which is punctuated by dramatic and surprising events. Things like the unanticipated discoveries of X rays or of radioactivity have taken place with remarkable frequency.

For example, in the 1930s, the neutron was discovered. So were the deuteron the positron, and the muon. And, let us not forget nuclear fission.

The 1940s saw the taming of the nuclear force, if such may be said of the use of the atomic bomb. The Lamb shift was measured, and it led to the surprising development of a consistent relativistic quantum theory. It was the decade of pions, and of the strange particle.

Parity-violation surprised almost everybody in the 1950s. So did the ever stranger properties of strange particles: associated production and neutral kaon behavior. Neutrinos were actually "seen" in the laboratory, and the first excited state of the nucleon was detected. An accelerator powerful enough to produce antimatter was commissioned. The discovery of the antiproton was a great accomplishment but hardly a surprise.

Schwinger was not surprised by the discovery of a second neutrino species in the 1960s, but most us were. The population explosion of new hadron species grew out of hand. As if by magic, these new particles filled out complete supermultiplets of the SU(3) symmetry scheme. Deep inelastic electron scattering produced convincing evidence for the existence of pointlike "partons" in the proton, which, like the atom, turned out not to be a plum pudding after all. And who predicted that time-reversal symmetry would be violated?

So it went in the 1970s. A kooky theory purporting to unify weak interactions and electromagnetism was shown to be renormalizable, and wary experimenters were amazed to find that its neutral currents were for real. The discovery of the J/ψ particle was a big surprise, and even the existence of charmed particles surprised some of us. As if that were not enough, there were the tau lepton, the upsilon particle and its associated beauty particles.

The 1980s has been a remarkably quiet decade. It was a surprise that CERN was able to mount a search for intermediate vector bosons in a timely and effective fashion. Any demonstration of international cooperation is surprising. However, the existence of W and Z bosons (like the existence of the antiproton) cannot really be thought of as something unexpected.

The lack of a fundamental but unanticipated discovery in this decade has not been for want of trying. Quite a number of surprises were *reported* during the 1980s. The trouble is that none of them seem really to be there. Perhaps a list of such nondiscoveries will suffice:

Magnetic monopole
Neutrino oscillations
Neutrino masses (twice: Russian and Canadian)
Zeta particle
No-neutrino double-beta decay
Muons from Cygnus X-3 (still alive?)
Proton decay
Forbidden decays like $\mu \rightarrow e\gamma$
Inexplicable "wrong sign dileptons"

Free quarks

Anomalons

About a half-dozen varieties of anomalous events seen at the CERN collider and purporting to show that there is new physics beyond the standard model. None of these effects is established. Those who have done so much to confirm the standard model have not, as yet, succeeded in demolishing it. Don't they wish!

What is the meaning of this almost incredible list of failed but noble efforts? Has the era of great surprises in particle physics come to an end? Have we exhausted nature's bag of tricks? Do we already have enough clues in hand to build a theory of everything? Or, have we set into effect a self-sustaining prophecy wherein no new discoveries will lead to no new machines, and a guarantee that there can be no discovery tomorrow? These are dangerous times for particle physics.

There are two approaches to our current dilemma, the possession of a theory which is on the one hand too successful, but on the other, clearly incomplete. There is the pedestrian and the grandiose: the upwards path from mere experiment to theory, and the downwards path of pure positive thinking: the way of Bohr, and the way of Einstein. I think that there *is* a lesson to be learned from the past. Bohr's route has proven itself to be successful beyond any reasonable expectation. Einstein's path—the search for a complete and unified theory *now*—has proven to be a dismal failure.

Some day, if our species lives so long, Einstein's dream may be fulfilled. Of course there *is* a connection between gravitation and the other forces of nature. Michael Faraday, like Einstein, and like all of us, believed in the existence of such a relation. Unlike Einstein, he was a follower of the upwards path. Towards the end of his life, on July 19, 1850, after an unsuccessful search for an experimentally verifiable connection between the forces, he wrote:

> Here end my trials for the present. The results are negative. They do not shake my strong feeling of the existence of a relation between gravity and electricity, though they give no proof that such a relation exists.

The Theory of Everything will come in its time if we let it. I am convinced that we still have a lot to learn about the phenomena of nature. One reason that Einstein failed in his quest is that he simply didn't know enough physics. Particle physics has not necessarily come to the end of the road. Astonishing experimental discoveries certainly remain to be made. If, and only if, we look, shall we find. The question is one of perseverance. Shall the scientific traditions established in the Renaissance survive in today's bizarrely materialistic society? The Way is clear, but what of the Will?

Big Things, Little Things

Elementary-particle physics—the search for the ultimate constituents of matter and the rules by which they combine—is concerned with the very smallest things. Cosmology—the study of the origin and the evolution of stars, galaxies, and of the universe itself—deals with the largest things of all. Between these two extremes, other sciences can be ranked in order of the size of the systems they deal with. The ladder of science leads upward from particle physics, to nuclear physics, atomic physics, chemistry, biology, geology, astronomy, and finally, to cosmology. Those who work at the ends of the ladder deal with the wildest possible extrapolations from everyday experience The cosmologist and the particle physicist would seem to have interests further apart than any other scientists.

And yes particle physics and cosmology have much in common. For both sciences, the basic rules are not yet set. Is the universe the unique product of the Big Bang, or are there an infinity of eternal and self-replicating universes? Are quarks and leptons the end of the line in the quest for the basic building blocks of matter, or are they (along with photons and other ''elementary'' particles) manifestations of an unbelievably tiny superstring?

No such grand uncertainties beset the other sciences. Chemistry may be chock-full of surprises (such as Teflon, silly putty, and napalm), but the underlying rules were all revealed in the first part of this century and almost certainly will never change. Biology, too, is full of wonders, most of them only dimly understood—the miracle of human awareness, for instance—but every biologist knows that life is essentially a chemical phenomenon, a product of electromagnetic interactions among complex aggregates of electrons and atomic nuclei. Cosmologists and particle physicists are the last prophets of radical revolution.

In recent decades, and particularly in the last one, the two extreme sciences have become inextricably linked. We are seeing the unification of the study of the immeasurably small with that of the inconceivably large. The marriage results from a fundamental tenet of science—that the laws of physics we learn on Earth are applicable throughout space and time. By studying atoms and particles on Earth, we have learned a lot about how stars are born, how they shine, and how they die; and we have begun to understand the origin of galaxies, of chemical elements, and of the universe.

Conversely, the observed properties of the universe tell us a great deal about particle physics. For example, the simple fact that the universe has not collapsed under its own gravity tells us that neutrinos, the most abundant particles, must be exceedingly light, if not altogether massless. In a more complex way, the observed abundance of helium in the universe constrains the number of neutrino species to be no greater than four. So far, physicists have seen two kinds of neutrinos, and they are fairly certain a third exists—there is one for each of the three known families of elementary particles. Cosmologists are telling us there can be at most one more neutrino and one more family, a prediction that will be tested at CERN's large accelerator in Geneva.

If one had to pick a year when macroverse and microverse began to converge, a good choice would be 1938. That was when Hans Bethe figured out that the Sun is powered by the fusion of atomic nuclei in its core—a result that spanned 23 powers of ten in size. Today we understand the fusion process in considerable detail, as well as how it governs the evolution of the Sun and other stars. It is a product of a complex interplay among all four known forces of nature.

Gravity makes the interior of the Sun a hellish place: the temperature is 15 million degrees, and the pressure is 100 trillion pounds per square inch. The intense heat and pressure strip electrons from atoms, allowing the bare nuclei (mostly protons, since the Sun is mostly made of hydrogen) to collide with one another. In a collision between two protons the weak force changes one proton into a neutron, which the strong force then binds to the other proton. After several more nuclear reactions the net result is that two protons and two neutrons are fused to form a helium nucleus. In the process a considerable quantity of energy is released. Electromagnetism, the fourth force, keeps the energy release under control—the positively charged protons repel one another, so that only a very few collisions actually result in fusion. Thanks to electromagnetism the Sun does not explode, but burns smoothly and reliably for billions of years.

A few billion years from now, though, once the Sun runs out of hydrogen, it will have to start fusing helium into larger elements, and that will cause it to swell into a red giant star, large enough to engulf Earth and the other inner planets. A dramatic end, perhaps, but the sun's fate is a mere whimper compared with the bang produced in the death throes of stars that start out much larger. Such a star manages the "hydrogen crisis" without difficulty; its internal temperature and pressure are so high that it can keep fusing heavier elements, until eventually it winds up with a core of iron. At that point fusion stops, and with no heat source to prop it up against gravity, the star implodes violently. About a Sun's worth of matter falls to the center and is compacted into a giant atomic nucleus: a neutron star. The collapse liberates an enormous amount of gravitational energy, causing the implosion to bounce into an explosion. The star goes supernova, becoming brighter than an entire galaxy.

This theory of stellar evolution has been extremely successful in explaining the life cycle of stars in terms of nuclear reactions. In recent years, however, it has had to endure not only triumph but also frustration at the hands of a new observational science: neutrino astronomy, which brings microphysics and macrophysics close together than ever.

The frustration came first with the work of Raymond Davis, who for almost 20 years has been patiently monitoring neutrinos coming from the Sun. Each time a proton in the solar core is converted into a neutron, two other particles are created: a positron, or antielectron, and a neutrino. The positron soon meets a free electron and annihilates, thereby releasing its energy, which diffuses slowly up to the solar surface and emerges as sunlight. The neutrino, on the other hand, passes right through the Sun. About 10 percent of the Sun's power is emitted in the form of these ghostly particles, many trillion of which pass harmlessly through your body every second of every day and every night.

In an enormous tank of cleaning fluid placed deep underground in a South Dakota gold mine, Davis has been managing to stop around a dozen solar neutrinos a month. That's the trouble—he is catching only a third as many neutrinos as the theory predicts he should. There is something strange going on inside the sun that we do not understand. It could be that there is something wrong with our solar model, a remote possibility being the existence of a black hole within the Sun. Another possibility is that there is something missing from our model of particle physics, and that neutrinos change their identity during their passage. This ''solar-neutrino problem'' is one of the big mysteries at the astroparticle frontier. Large experiments are being launched in Russia, Canada, Japan, and Italy to study the Sun's neutrinos, and to find out just what is happening deep inside the Sun.

More recently, the theory of stellar evolution scored a major triumph. In February 1987, astronomers and other sky-watchers saw something that had not been seen on Earth for four centuries: a nearby supernova. What wonders it revealed! And most wonderful of all, as far as physicists were concerned, was the wave of neutrinos that burst out of the dying star. Just as astrophysicists had predicted, the supernova we saw—and Supernova 1987a was clearly visible to the naked eye—was but a tiny part of the action. About 99 percent of the explosion's energy was emitted in the form of neutrinos, which made their way out of the star in just a few seconds, while the photons were still struggling to escape.

The fact that a few of the supernova's neutrinos were detected on earth was an incredible act of serendipity, and a perfect illustration of how particle physics and astrophysics have come together. In the early eighties, years before Supernova 1987a was sighted, large underground detectors had been deployed in Japan and in Ohio. To watch for supernovas? Certainly not. Their purpose had been to search for evidence that protons decay spontaneously and that all earthly matter is therefore mortal. Particle physicists had been so proud of their successful unification of the weak and electromagnetic forces that they had put forward a grand unification of all of the forces of nature except gravity. The simplest such model, advanced by Howard Georgi and me, predicted that the average proton would live 10^{30} years. By closely watching a test mass of a lot more than 10^{30} protons, one should thus be able to see a few of them decay.

Unfortunately, the experiments have failed so far to see proton decay—my theories did not pan out. But they did see about 20 of the neutrinos from the supernova, thereby proving that astrophysicists really do understand what a supernova is.

What's more, the observations revealed something about neutrinos as well. All the neutrinos from the supernova arrived at the detectors within a 10 second interval. That tells us they were all traveling at nearly the same speed, practically the speed of light. It follows that the detected neutrinos—which were of the type known as electron neutrinos—are almost massless. In fact, the upper limit on the electron neutrino's mass that has been calculated from observations of the supernova is about as good as the one derived from three decades of laboratory measurements.

That result bears on another central issue in contemporary astro-cum-particle physics. About a decade ago, after centuries of training their telescopes on everything from lowly clouds of gas and dust to stars, galaxies, and galaxy clusters, astronomers made a startling and (to them) disconcerting discovery: they found that most of the matter of the universe is "none of the above." It is not in the form of luminous stars or groups of stars. Nor is it in the form of gas or dust, which would reveal itself by the absorption of light from distant stars. By measuring the rotation velocities of galaxies, astronomers have shown that galaxies spin much faster than they would if luminous matter was all they were made of. Instead, 80 to 90 percent of a galaxy's mass seems to be stored in a vast spherical halo made of mysterious, dark, and invisible matter.

What is this stuff? One possibility is that dark matter consists of neutrinos. This idea could work if there were a neutrino species with a mass of about a ten-thousandth of that of the electron. It can't be the electron neutrino—the supernova has confirmed that it is too light. While we know of two other neutrino species, most cosmologists have concluded that neutrinos do not quite fit the bill. Computer simulations of galaxy formation suggest that massive neutrinos would lead to a universe quite different from the one we inhabit. Consequently, the dark matter of the universe is likely to be something else, something lying outside today's particle bestiary.

Here is where particle physicists and cosmologists get really intimate. The universe is telling us that there must be some new kind of particle or thing constituting the bulk of its mass. it looks like the ancient Greeks were right again! They told us that earthly matter was made up of four elements, and that heavenly matter consists of a fifth and perfect element. Fire, water, earth, and air have already been replaced by protons, neutrons, electrons, and neutrinos, and we are now hot on the trail of the cosmic quintessence.

Many suggestions have been put forward as to what the dark stuff is: axions, photinos, magnetic monopoles, and weakly interacting massive particles (WIMPs) are a few possibilities. My friends and I have our own favorite candidate: charged massive particles, or CHAMPs. If we are right, this new form of matter is disguised as superheavy isotopes of ordinary chemical elements: half of it as hydrogen with an atomic weight of a few hundred thousand, and the other half (the anti-CHAMPs) as superheavy isotopes of other elements. Because they are so very heavy, these particles don't lose their energy when they collide with other particles in interstellar space, anymore than an eighteen-wheeler loses energy when flies hit its windshield. Because they cannot slow down, they cannot collapse to form stars, and so they remain dark.

But they do bombard Earth at a rate of about one per square inch per second. Once in the atmosphere, CHAMPs form molecules of superheavy water that rain down to the surface of Earth. After accumulating, molecule by molecule, for billions of years, they should today constitute some ten parts per billion by weight of earth's crust: they should be more abundant than gold and somewhat less than silver.

A group of researchers at Harvard has begun to search for CHAMPs in earnest. They begin with such a promising sample of matter as deep-sea sediment. They extract water from the sample and subject it to ultracentrifugation, thus concentrating CHAMPs, if they exist, into a tiny droplet. They then measure the density of the droplet to see if it is higher than one would expect in the absence of CHAMPs. It is an easy enough experiment, but so far as we know, no one has done it before. The idea of CHAMPs as dark matter seems perfectly consistent with everything we know, but it is a little bit crazy. Probably, scientists will fail to find any CHAMPs for the simple reason that they don't exist.

But just suppose that our whim is nature's way and that we succeed in finding these particles. If they really exist, it would be child's play to extract them in bulk. Think of the things we could do with them. Aside from all the wonderful new materials we could fabricate, we could separate CHAMPs from anti-CHAMPs and annihilate them in an accelerator. Because they are so massive, we wouldn't need to accelerate them much to get high energies. In other words, CHAMPs could offer a cheap way to study high-energy physics—and perhaps also to produce abundant, usable energy in a safe way. They may even be the fuel we need to send spaceships to the stars. I hope we find them.

High-energy physics, with or without CHAMPs, will continue to be well worth studying. Over the past few decades, it has been our ticket to a frontier more distant than the stars, a frontier of the human imagination. It is there that the real excitement in physics and cosmology lies: in trying to understand the mysterious mechanism that led to the Big Bang 15 billion years ago. Today our universe is deathly cold—at three degrees above absolute zero, it is about the temperature of liquid helium. (Of course, it isn't all that cold here in the cozy neighborhood of our sun.) When the universe was a few hundred thousand years old, however, it was uniformly hot—about as hot as a welding torch. That's when atoms first formed. Earlier still, when the universe was just a few minutes old, it was even hotter, around 10 billion degrees, which corresponds to particles moving with energies of a million electron volts. We understand the early history of the universe only because we have studied nuclear reactions at these energies in accelerators. For instance, William Fowler and his collaborators at the California Institute of Technology have shown that most of the small atomic nuclei in the universe, such as helium, deuterium, and lithium, were produced in the first few minutes.

In the past few years, the large proton-antiproton colliders at CERN in Geneva and at Fermilab in Illinois have been exploring phenomena at trillions of electron volts. Those energies correspond to the temperature of the universe when it was just a few trillionths of a second old. So far, the experiments have confirmed the theory that unifies the electromagnetic and the weak force. Otherwise, nothing very exciting has shown up. The most exciting developments, it now appears, took place a bit earlier.

The 52-mile-long Superconducting Super Collider, if it is built in Texas as now planned, may reveal some answers to our questions. It will push the high-energy frontier upward by another factor of 20. When the universe was very young and burning bright, all the particles we know were massless, and the symmetry between weak and electromagnetic forces was perfect. It was a very different universe from the cold and disordered one we now inhabit. Somehow, the symmetry was broken, the weak force became weak, and the various quarks and leptons and weak bosons acquired the seemingly random masses they now have. How did all this come about? By providing a tiny window upon the universe at the very moment it crystallized, the Super Collider may help solve the riddle of the origin of mass.

But not even the Super Collider, nor its successors, will approach the ultimate question of the origin of the universe. To recreate the temperature of the universe at its birth we would need an accelerator light years in size. That does not mean, however, that we can't ever learn the nature of the infant universe. We have a tool that is surely up to the task: the human brain.

Indeed, as we near the end of a decade of remarkable progress, many of my colleagues and I have a strong feeling that we are on the verge of a profound new synthesis, one that can fulfill our sacred duty to understand, as best we can, the universe into which we were born. The hope is alive in us that one day soon, through the power of pure thought (with a lot of help from experimental physicists and astronomers), our species will begin to understand just how and why the universe came to be, and what it will do long after human beings are extinct.

Author's note: Astronomers and physicists have pretty well ruled out the CHAMP hypothesis. No evidence of their existence has shown up on Earth or in the stars.

THE WORK OF A THEORIST: GRAND UNIFICATION

Towards a Unified Theory: Threads in a Tapestry

In 1956, when I began doing theoretical physics, the study of elementary particles was like a patchwork quilt. Electrodynamics, weak interactions, and strong interactions were clearly separate disciplines, separately taught and separately studied. There was no coherent theory that described them all. Developments such as the observation of parity violation, the successes of quantum electrodynamics, the discovery of hadron resonances and the appearance of strangeness were well-defined parts of the picture, but they could not be easily fitted together.

Things have changed. Today we have what has been called a ''standard theory'' of elementary-particle physics in which strong, weak, and electromagnetic interactions all arise from a local symmetry principle. It is, in a sense, a complete and apparently correct theory, offering a qualitative description of all particle phenomena and precise quantitative predictions in many instances. There is no experimental data that contradicts the theory. In principle, if not yet in practice, all experimental data can be expressed in terms of a small number of ''fundamental'' masses and coupling constants. The theory we now have is an integral work of art: the patchwork quilt has become a tapestry.

Tapestries are made by many artisans working together. The contributions of separate workers cannot be discerned in the completed work, and the loose and false threads have been covered over. So it is in our picture of particle physics. Part of the picture is the unification of weak and electromagnetic interactions and the prediction of neutral currents, celebrated by the award of the Nobel Prize. Another part concerns the reasoned evolution of the quark hypothesis from mere whimsy to established dogma. Yet another is the development of quantum chromodynamics into a plausible, powerful, and predictive theory of strong interactions. All is woven together in the tapestry; one part makes little sense without the other. Even the development of the electroweak theory was not as simple and straightforward as it might have been. It did not arise full blown in the mind of one physicist, nor even of three. It, too, is the result of the collective endeavor of many scientists, both experimenters and theorists.

Let me stress that I do not believe that the standard theory will long survive as a correct and complete picture of physics. All interactions may be gauge interactions, but surely they must lie within a unifying group. This would imply the existence of a new and very weak interaction which mediates the decay of protons. All matter is thus inherently unstable, and can be observed to decay. Such a synthesis of weak, strong, and electromagnetic interactions has been called a ''grand unified theory,'' but a theory is neither grand nor unified unless it includes a description of gravitational phenomena. We are still far from Einstein's truly grand design.

Physics of the past century has been characterized by frequent great but unanticipated experimental discoveries. If the standard theory is correct, this age has come to an end. Only a few important particles remain to be discovered, and many of their properties are alleged to be known in advance. Surely this is not the way things will be, for nature must still have some surprises in store for us.

Nevertheless, the standard theory will prove useful for years to come. The confusion of the past is now replaced by a simple and elegant synthesis. The standard theory may survive as a part of the ultimate theory, or it may turn out to be fundamentally wrong. In either case, it will have been an important way station, and the next theory will have to be better.

In this essay, I shall not attempt to describe the tapestry as a whole, nor even that portion which is the electroweak synthesis and its empirical triumph. Rather, I shall describe several old threads, mostly overwoven, which are closely related to my own researches. My purpose is not so much to explain who did what when, but to approach the more difficult question of why things went as they did. I shall also follow several new threads which may suggest the future development of the tapestry.

Early Models

In the 1920s, it was still believed that there were only two fundamental forces: gravity and electromagnetism. In attempting to unify them, Einstein might have hoped to formulate a universal theory of physics. However, the study of the atomic nucleus soon revealed the need for two additional forces: the strong force to hold the nucleus together and the weak force to enable it to decay. Yukawa asked whether there might be a deep analogy between these new forces and electromagnetism. All forces, he said, were to result from the exchange of mesons. His conjectured mesons were originally intended to mediate both the strong and the weak interactions: they were strongly coupled to nucleons and weakly coupled to leptons. This first attempt to unify strong and weak interactions was fully 40 years premature. Not only this, but Yukawa could have predicted the existence of neutral currents. His neutral meson, essential to provide the charge independence of nuclear forces, was also weakly coupled to pairs of leptons.

Not only is electromagnetism mediated by photons, but it arises from the requirement of local gauge invariance. This concept was generalized in 1954 to apply to non-Abelian local symmetry groups.[1] It soon became clear that a more far-reaching analogy might exist between electromagnetism and the other forces. They, too, might emerge from a gauge principle.

At bit of a problem arises at this point. All gauge mesons must be massless, yet the photon is the only massless meson. How do the other gauge bosons get their masses? There was no good answer to this question until the work of Weinberg and Salam[2] as proven by 't Hooft[3] (for spontaneously broken gauge theories) and of Gross, Wilczek, and Politzer[4] (for unbroken gauge theories). Until this work was done, gauge meson masses had simply to be put in *ad hoc*.

Sakurai suggested in 1960 that strong interactions should arise from a gauge principle.[5] Applying the Yang–Mills construct to the isospin-hypercharge symmetry group, he predicted the existence of the vector mesons ρ and ω. This was the first phenomenological $SU(2) \times U(1)$ gauge theory. It was extended to local SU(3) by Gell-Mann and Ne'eman in 1961.[6] Yet, these early attempts to formulate a gauge theory of strong interactions were doomed to fail. In today's jargon, they used "flavor" as the relevant dynamical variable, rather than the hidden and then unknown variable "color." Nevertheless, this work prepared the way for the emergence of quantum chromodynamics a decade later.

Early work in nuclear beta decay seemed to show that the relevant interaction was a mixture of S, T, and P. Only after the discovery of parity violation, and the undoing of several wrong experiments, did it become clear that the weak interactions were in reality VA. The synthesis of Feynman and Gell-Mann and of Marshak and Sudarshan was a necessary precursor to the notion of a gauge theory of weak interactions.[7] Bludman formulated the first SU(2) gauge theory of weak interactions in 1958.[8] No attempt was made to include electromagnetism. The model included the conventional charged-current interactions, and in addition, a set of neutral current couplings. These are of the same strength and form as those of today's theory in the limit in which the weak mixing angle vanishes. Of course, a gauge theory of weak interactions alone cannot be made renormalizable. For this, the weak and electromagnetic interactions must be unified.

Schwinger, as early as 1956, believed that the weak and electromagnetic interactions should be combined together into a gauge theory.[9] The charged massive vector intermediary and the massless photon were to be the gauge mesons. As his student, I accepted this faith. In my 1958 Harvard thesis, I wrote: "It is of little value to have a potentially renormalizable theory of beta processes without the possibility of a renormalizable electrodynamics. We should care to suggest that a fully acceptable theory of these interactions may only be achieved if they are treated together . . . "[10] We used the original SU(2) gauge interaction of Yang and Mills. Things had to be arranged so that the charged current, but not the neutral (electromagnetic) current, would violate parity and strangeness. Such a theory is technically possible to construct, but it is both ugly and experimentally false.[11] We know now that neutral currents do exist and that the electroweak gauge group must be larger than SU(2).

Another electroweak synthesis without neutral currents was put forward by Salam and Ward in 1959.[12] Again, they failed to see how to incorporate the experimental fact of parity violation. Incidentally, in a continuation of their work in 1961, they suggested a gauge theory of strong, weak, and electromagnetic interactions based on the local symmetry group SU(2) × SU(2).[13] This was a remarkable portent of the SU(3) × SU(2) × U(1) model which is accepted today.

We come to my own work[14] done in Copenhagen in 1960, and done independently by Salam and Ward.[15] We finally saw that a gauge group larger than SU(2) was necessary to describe the electroweak interactions. Salam and Ward were motivated by the compelling beauty of gauge theory. I thought I saw a way to a renormalizable scheme. I was led to the group SU(2) × U(1) by analogy with the approximate isospin-hypercharge group which characterizes strong interactions. In this model there were two electrically neutral intermediaries: the massless photon and a massive neutral vector meson which I called B but which is now known as Z. The weak mixing angle determined to what linear combination of SU(2) × U(1) generators B would correspond. The precise form of the predicted neutral-current interaction has been verified by recent experimental data. However, the strength of the neutral current was not prescribed, and the model was not in fact renormalizable. These glaring omissions were to be rectified by the work of Salam and Weinberg and the subsequent proof of renormalizability. Furthermore, the model was a model of leptons—it could not evidently be extended to deal with hadrons.

Renormalizability

In the late fifties, quantum electrodynamics and pseudoscalar meson theory were known to be renormalizable, thanks in part to work of Salam. Neither of the customary models of weak interactions—charged intermediate vector bosons or direct four-fermion couplings—satisfied this essential criterion. My thesis at Harvard, under the direction of Julian Schwinger, was to pursue my teacher's belief in a unified electroweak gauge theory. I had found some reason to believe that such a theory was less singular than its alternatives. Feinberg, working with charged intermediate vector mesons, discovered that a certain type of divergence would cancel for a special value of the meson anomalous magnetic moment.[16] It did not correspond to a "minimal electromagnetic coupling," but to the magnetic properties demanded by a gauge theory. Tzou Kuo-Hsien examined the zero-mass limit of charged vector meson electrodynamics.[17] Again, a sensible result is obtained only for a very special choice of the magnetic dipole moment and electric quadrupole moment, just the values assumed in a gauge theory. Was it just coincidence that the electromagnetism of a charged vector meson was least pathological in a gauge theory?

Inspired by these special properties, I wrote a notorious paper.[18] I alleged that a softly-broken gauge theory, with symmetry breaking provided by explicit mass terms, was renormalizable. It was quickly shown that this is false.

Again, in 1970, Iliopoulos and I showed that a wide class of divergences that might be expected would cancel in such a gauge theory.[19] We showed that the naive divergences of order $(\alpha\Lambda^4)^n$ were reduced to "merely" $(\alpha\Lambda^2)^n$, where Λ is a cutoff momentum. This is probably the most difficult theorem that Iliopoulos or I had even proven. Yet, our labors were in vain. In the spring of 1971, Veltman informed us that his student Gerhart 't Hooft had established the renormalizability of spontaneously broken gauge theory.

In pursuit of renormalizability, I had worked diligently but I completely missed the boat. The gauge symmetry is an exact symmetry, but it is hidden. One must not put in mass terms by hand. The key to the problem is the idea of spontaneous symmetry breakdown: the work of Goldstone as extended to gauge theories by Higgs and Kibble in 1964.[20] These workers never thought to apply their work on formal field theory to a phenomenologically relevant model. I had had many conversations with Goldstone and Higgs in 1960. Did I neglect to tell them about my SU(2) × U(1) model, or did they simply forget?

Both Salam and Weinberg had had considerable experience in formal field theory, and they had both collaborated with Goldstone on spontaneous symmetry breaking. In retrospect, it is not so surprising that it was they who first used the key. Their SU(2) × U(1) gauge symmetry was spontaneously broken. The masses of the W and Z and the nature of neutral current effects depend on a single measurable parameter, not two as in my unrenormalizable model. The strength of the neutral currents was correctly predicted. The daring Weinberg-Salam conjecture of renormalizability was proven in 1971. Neutral currents were discovered in 1973,[21] but not until 1978 was it clear that they had just the predicted properties.[22]

The Strangeness-Changing Neutral Current

I had more or less abandoned the idea of an electroweak gauge theory during the period 1961–1970. Of the several reasons for this, one was the failure of my naive foray into renormalizability. Another was the emergence of an empirically successful description of strong interactions—the SU(3) unitary symmetry scheme of Gell-Mann and Ne'eman. This theory was originally phrased as a gauge theory, with ρ, ω, and K* as gauge mesons. It was completely impossible to imagine how both strong and weak interactions could be gauge theories: there simply wasn't room enough for commuting structures of weak and strong currents. Who could foresee the success of the quark model, and the displacement of SU(3) from the arena of flavor to that of color? The predictions of unitary symmetry were being borne out—the predicted Ω^- was discovered in 1964. Current algebra was being successfully exploited. Strong interactions dominated the scene.

When I came upon the SU(2) × U(1) model in 1960, I had speculated on a possible extension to include hadrons. To construct a model of leptons alone seemed senseless: nuclear beta decay, after all, was the first and foremost problem. One thing seemed clear. The fact that the charged current violated strangeness would force the neutral current to violate strangeness as well. It was already well known that strangeness-changing neutral currents were either strongly suppressed or absent. I concluded that the Z^0 had to be made very much heavier than the W. This was an arbitrary but permissible act in those days: the symmetry-breaking mechanism was unknown. I had "solved" the problem of strangeness-changing neutral currents by suppressing all neutral currents: the baby was lost with the bath water.

I returned briefly to the question of gauge theories of weak interactions in a collaboration with Gell-Mann in 1961.[23] From the recently developing ideas of current algebra we showed that a gauge theory of weak interactions would inevitably run into the problem of strangeness-changing neutral currents. We concluded that something essential was missing. Indeed it was. Only after quarks were invented could the idea of the fourth quark and the Glashow-Iliopoulos-Maiani (GIM) mechanism arise.

From 1961 to 1964, Sidney Coleman and I devoted ourselves to the exploitation of the unitary symmetry scheme. In the spring of 1964, I spent a short leave of absence in Copenhagen. There, Bjorken and I suggested that the Gell-Mann—Zweig-system of three quarks should be extended to four.[24] (Other workers had the same idea at the same time.[25]) We called the fourth quark the charmed quark. Part of our motivation for introducing a fourth quark was based on our mistaken notions of hadron spectroscopy. But we also wished to enforce an analogy between the weak leptonic current and the weak hadronic current. Because there were two weak doublets of leptons, we believed there had to be two weak doublets of quarks as well. The basic idea was correct, but today there seem to be three doublets of quarks and three doublets of leptons.

The weak current Bjorken and I introduced in 1964 was precisely the GIM current. The associated neutral current, as we noted, conserved strangeness. Had we inserted these currents into the earlier electroweak theory, we would have solved the problem of strangeness-changing neutral currents. We did not. I had apparently quite forgotten my earlier ideas of electroweak synthesis. The problem which was explicitly posed in 1961 was solved, in principle, in 1964. No one, least of all me, knew it. Perhaps we were all befuddled by the chimera of relativistic SU(6), which arose at about this time to cloud the minds of theorists.

Five years later, John Iliopoulos, Luciano Maiani and I returned to the question of strangeness-changing neutral currents.[26] It seems incredible that the problem was totally ignored for so long. We argued that unobserved effects (a large K_1K_2 mass difference; decays like $K \rightarrow \pi\nu\bar{\nu}$; etc.) would be expected to arise in any of the known weak interaction models: four fermion couplings; charged vector meson models; or the electroweak gauge theory. We worked in terms of cutoffs, since no renormalizable theory was known at the time. We showed how the unwanted effects would be eliminated with the conjectured existence of a fourth quark. After languishing for a decade, the problem of the selection rules of the neutral current was finally solved. Of course, not everyone believed in the predicted existence of charmed hadrons.

This work was done fully three years after the epochal work of Weinberg and Salam, and was presented in seminars at Harvard and at MIT. Neither I, nor my co-workers, nor Weinberg, sensed the connection between the two endeavors. We did not refer, nor were we asked to refer, to the Weinberg–Salam work in our paper.

The relevance became evident only a year later. Due to the work of 't Hooft, Veltman, Lee, and Zinn-Justin, it became clear that the Weinberg-Salam *ansatz* was in fact a renormalizable theory. With GIM, it was trivially extended from a model of leptons to a theory of weak interactions. The ball was now squarely in the hands of the experimenters. Within a few years, charmed hadrons and neutral currents were discovered, and both had just the properties they were predicted to have.

From Accelerators to Mines

Pions and strange particles were discovered by passive experiments which made use of the natural flux of cosmic rays. However, in the last three decades, most discoveries in particle physics were made in the active mode, with the artificial aid of particle accelerators. Passive experimentation stagnates from a lack of funding and lack of interest. Recent developments in theoretical particle physics and in astrophysics may mark an imminent rebirth of passive experimentation. The concentration of virtually all high-energy physics endeavors at a small number of major accelerator laboratories may be a thing of the past.

This is not to say that the large accelerator is becoming extinct; it will remain an essential if not exclusive tool of high-energy physics. Do not forget that the existence of Z^0 at ~100 GeV is an essential but quite untested prediction of the electroweak theory. There will be additional dramatic discoveries at accelerators, and these will not always have been predicted in advance by theorists. The construction of new machines like LEP and ISABELLE is mandatory.

Consider the successes of the electroweak synthesis, and the fact that the only plausible theory of strong interactions is also a gauge theory. We must believe in the ultimate synthesis of strong, weak, and electromagnetic interactions. It has been shown how the strong and electroweak gauge groups may be put into a larger but simple gauge group.[27] Grand unification—perhaps along the lines of the original SU(5) theory of Georgi and me—must be essentially correct. This implies that the proton, and indeed all nuclear matter, must be inherently unstable. Sensitive searches for proton decay are now being launched. If the proton lifetime is shorter than 10^{32} years, as theoretical estimates indicate, it will not be long before it is seen to decay.

Once the effect is discovered (and I am sure it will be), further experiments will have to be done to establish the precise modes of decay of nucleons. The selection rules, mixing angles, and space-time structure of a new class of effective four-fermion couplings must be established. The heroic days of the discovery of the nature of beta decay will be repeated.

The first generation of proton decay experiments is cheap, but subsequent generations will not be. Active and passive experiments will compete for the same dwindling resources.

Other new physics may show up in elaborate passive experiments. Today's theories suggest modes of proton decay which violate both baryon number and lepton number by unity. Perhaps this $\Delta B = \Delta L = 1$ law will be satisfied. Perhaps $\Delta B = -\Delta L$ transitions will be seen. Perhaps, as Pati and Salam suggest, the proton will decay into three leptons. Perhaps two nucleons will annihilate in $\Delta B = 2$ transitions. The effects of neutrino oscillations resulting from neutrino masses of a fraction of an election volt may be detectable. ''Superheavy isotopes'' which may be present in the Earth's crust in small concentrations could reveal themselves through their multi-GeV decays. Neutrino bursts arising from distant astronomical catastrophes may be seen. The list may be endless or empty. Large passive experiments of the sort now envisioned have never been done before. Who can say what results they may yield?

Premature Orthodoxy

The discovery of the J/Ψ in 1974 made it possible to believe in a system involving just four quarks and four leptons. Very quickly after this a third charged lepton (the tau) was discovered, and evidence appeared for a third $Q = -1/3$ quark (the b quark). Both discoveries were classic surprises. It became immediately fashionable to put the known fermions into families or generations:

$$\begin{pmatrix} u & \nu_e \\ d & e \end{pmatrix} \begin{pmatrix} c & \nu_\mu \\ s & \mu \end{pmatrix} \begin{pmatrix} t & \nu_\tau \\ b & \tau \end{pmatrix}$$

The existence of a third $Q = 2/3$ quark (the t quark) is predicted. The Cabibbo-GIM scheme is extended to a system of six quarks. The three family system is the basis to a vast and daring theoretical endeavor. For example, a variety of papers have been written putting experimental constraints on the four parameters which replace the Cabibbo angle in a six-quark system. The detailed manner of decay of particles containing a single b quark has been worked out. All that is wanting is experimental confirmation. A new orthodoxy has emerged, one for which there is little evidence, and one in which I have little faith.

This is not the place to describe our views in detail. They are very speculative and probably false. The point I wish to make is simply that it is too early to convince ourselves that we know the future of particle physics. There are too many points at which the conventional picture may be wrong or incomplete. The $SU(3) \times SU(2) \times U(1)$ gauge theory with three families is certainly a good beginning, not to accept but to attack, extend, and exploit. We are far from the end.

References

1. Yang, C. N. and Mills, R., Phys. Rev. *96*, 191 (1954). Also, Shaw, R., unpublished.

2. Weinberg, S., Phys. Rev. Lett. *19,* 1264 (1967); Salam, A., in *Elementary Particle Physics* (ed. Svartholm, N.; Almqvist and Wiksell; Stockholm; 1968).
3. 't Hooft, G., Nuclear Physics *B 33,* 173; *35,* 167 (1971); Lee, B. W., and Zinn-Justin, J., Phys. Rev. D *5,* 3121–3160 (1972); 't Hooft, G., and Veltman M., Nuclear Physics *B 44,* 189 (1972).
4. Gross, D. J. and Wilczek, F., Phys. Rev. Lett. *30,* 1343 (1973); Politzer, H. D., Phys. Rev. Lett. *30,* 1346 (1973).
5. Sakurai, J. J., Annals of Physics *11,* 1 (1960).
6. Gell-Mann, M., and Ne'eman, Y., *The Eightfold Way* (Benjamin, W. A., New York, 1964).
7. Feynman, R., and Gell-Mann, M., Phys. Rev. *109,* 193 (1958); Marshak, R., and Sudarshan, E. C. G., Phys. Rev. *109,* 1860 (1958).
8. Bludman, S., Nuovo Cimento Ser. 10 *9,* 433 (1958).
9. Schwinger, J., Annals of Physics *2,* 407 (1958).
10. Glashow, S. L., Harvard University thesis, p. 75 (1958).
11. Georgi, H., and Glashow, S. L., Phys. Rev. Lett. *28,* 1494 (1972).
12. Salam, A., and Ward, J., Nuovo Cimento *11,* 568 (1959).
13. Salam, A., and Ward, J., Nuovo Cimento *19,* 165 (1961).
14. Glashow, S. L., Nuclear Physics *22,* 579 (1961).
15. Salam, A., and Ward, J., Physics Lett. *13,* 168 (1964).
16. Feinberg, G., Phys. Rev. *110,* 1482 (1958).
17. Tzou Kuo-Hsien, Comptes Rendus *245,* 289 (1957).
18. Glashow, S. L., Nuclear Physics *10,* 107 (1959).
19. Glashow, S. L., and Iliopoulos J., Phys. Rev. D *3,* 1043 (1971).
20. Many authors are involved with this work: Brout, R., Englert, F., Goldstone, J., Guralnik, G., Hagen, C., Higgs, P., Jona-Lasinio, G., Kibble, T., and Nambu, Y.
21. Hasert, F. J., *et al.,* Physics Letters *46 B,* 138 (1973); Nuclear Physics *B 73,* 1 (1974); Benvenuti, A., *et al.,* Phys. Rev. Lett. *32,* 800 (1974).
22. Prescott, C. Y., *et al.,* Phys. Lett. *B 77,* 347 (1978).
23. Gell-Mann, M., and Glashow, S. L., Annals of Physics *15,* 437 (1961).
24. Bjorken, J., and Glashow, S. L., Physics Letters *11,* 84 (1964).
25. Amati, D., *et al.,* Nuovo Cimento *34,* 1732 (A 64); Hara, Y. Phys. Rev. *134,* B701 (1964); Okun, L. B., Phys. Lett. *12,* 250 (1964); Maki, Z., and Ohnuki, Y., Prog. Theor. Phys. *32,* 144 (1964); Nauenberg, M., (unpublished); Teplitz, V., and Tarjanne, P., Phys. Rev. Lett. *11,* 447 (1963).
26. Glashow, S. L., Iliopoulos, J., and Maiani, L., Phys. Rev. D *2,* 1285 (1970).
27. Georgi, H., and Glashow, S. L., Phys. Rev. Lett. *33,* 438 (1974).

Unified Theory of Elementary-Particle Forces

WITH HOWARD GEORGI

The notion of what are the "elementary" or structureless constituents of matter keeps changing as we are able to probe smaller and smaller distances with higher and higher energies. As long as we were limited by the energy available in chemical processes, the elementary particles were atoms; later they were protons, neutrons and electrons; currently we can smash matter into pieces sufficiently fine that quarks and leptons appear to be the elementary constituents of matter.

Quarks are the constituents of the hadrons, the particles (such as protons or pions) that interact via the strong force. There is evidence for at least five types ("flavors") of quarks; they are generally labelled u (for up) d (for down), s (for strange), c (for charm) and b (for bottom or beauty). Further, each type of quark comes in three varieties ("colors"), generally labelled 'red," "white" (or sometimes "green") and "blue." The strong interaction, which binds quarks into hadrons, is (we believe) an interaction in which the three colors play a role analogous to the charge in electrodynamics. The currently accepted theory of these interactions is quantum chromodynamics, or QCD for short.

At this time there is some evidence for six kinds of leptons: the electron, the muon and the tau, and the neutrinos associated with each. The leptons have no color and are almost completely unaffected by the chromodynamic interactions.

Except for the three neutrinos, quarks and leptons carry electric charges, and therefore interact electromagnetically. All the particles, including neutrinos, interact via the weak interaction. Particles that interact electromagnetically do not change their identity: except for possible particle-antiparticle pairs, the particles that emerge after an interaction have the same properties (charge, strangeness, and so on) as the particles that entered into the interaction. The weak interactions, on the other hand, do change particle identities. The classic example is beta decay in which a neutron decays into a proton, an electron and an antineutrino ($\overline{\nu}_e$); this occurs when one of the d quarks in the neutron decays into a u (which remains bound to the other d and u that were in the neutron) together with an electron and an antineutrino.

Unification

Quantum electrodynamics, quantum chromodynamics, and the currently accepted theory of weak interactions adequately describe the forces among the elementary particles down to 10^{-15} cm (about 1 percent of the proton radius), which is the shortest distance probed by today's accelerators. What happens at shorter distances we do not know. But we suspect that at distances of the order of 10^{-29} cm, all three interactions—along with others not yet observed—will be unified. That is, all interactions will have the same strength and the distinctions between quarks, antiquarks and leptons will disappear.

Although this proposed unification takes place at ridiculously small distances (or high energies), it has important consequences for the "low-energy" world of contemporary physics. One such consequence is a prediction of an angle that appears in the theory of weak interactions, called θ_w, the weak mixing angle. Unified theories predict $\sin^2\theta_w \cong .20$. When this prediction was first worked out, the best experimental value was $\sin^2\theta_w \cong .35$. But as the experiments have improved, the value has marched steadily down; the experimental value now is $\sin^2\theta_w = .23 \pm .02$, almost in agreement with the unified theory.

The most spectacular consequence of the unification has yet to be tested conclusively. It is that the proton itself (and thus all matter) is unstable: In a time on the order of 10^{31} years a proton can decay into a positron and a π^0. The positron eventually annihilates an electron, producting gamma rays, and the π^0 decays quickly into gamma rays as well. The net result is that a hydrogen atom disappears into energy. Although the chance that any one proton decays in any one year is absurdly small, a macroscopic chunk of material has a very large number of protons. Experiments are now in preparation to look at very large samples of material to test the prediction.

Before explaining the unification we briefly review its components, QCD, quantum electrodynamics and the weak interactions, in a language particularly suited to the unification, the language of gauge theories. For our examples we will chiefly use the quarks u and d and the leptons e and ν. These are the constituents of all ordinary materials; the heavier quarks and leptons are of interest almost exclusively to particle physicists.

Electrodynamics

Quantum electrodynamics (QED) is the quantum theory of the electromagnetic interactions of charged point particles. It is an extremely successful theory in that its predictions have been verified experimentally to great precision, in part because it is characterized by a small dimensionless coupling constant. The concept of a "coupling constant" will be central to our idea of unification, so we will describe it in some detail.

The idea starts with Cuolomb's law. In relativistic quantum theory it is natural to measure the products of charges (in electrostatic units) in units of $\hbar c$; it is also natural to make use of the fact that charges are quantized as integral multiples of the proton charge, e. Thus we write Coulomb's law as

$$F/\hbar c = \alpha QQ'/r^2$$

where Q and Q' are integers. We have absorbed the unit of charge, e, into the dimensionless coupling constant

$$\alpha = e^2/\hbar c$$

whose experimental value is pretty nearly $1/137$. It is small enough to allow straightforward perturbation expansions in powers of a.

We should emphasize that while the quantization of electromagnetic charge is an experimental fact, quantum electrodynamics does not require it. The theory would still make sense if there were also a particle with electric charge $Q = \sqrt{2}$, for example. But in the unified theory we will describe below, charge quantization is automatic. Also note that the charge Q is an additive quantum number, and that it has opposite values for a particle and its antiparticle.

All the theories we will discuss are relativistic quantum field theories. In such a theory both the forces that act on particles and the particles themselves are represented by quantized fields. A field and its corresponding quanta represent dual aspects of matter: A field quantum behaves as a particle and the expectation value of the field is the wave function of the particle. We shall distinguish between the "particles" that correspond to the constituents of matter (these are the quarks and leptons—particles with spin ½) and "forces" that determine the interactions between particles (these are associated with particles of integral spin, such as the photon).

The dynamics of the electromagnetic force are given by the vector potential, $\mathbf{A}(x)$—the "photon field." Quantum electrodynamics is invariant under an extremely large group of transformations. We can add to the vector potential the gradient of any scalar function and simultaneously change the phase of all particle fields by an amount proportional to the scalar:

$$\mathbf{A}(x) \rightarrow \mathbf{A}(x) + \nabla\theta(x)$$
$$\psi(x) \rightarrow \psi(x)e^{ieQ\theta(x)/\hbar c}$$

and this transformation leaves the theory unchanged. This symmetry is called a gauge invariance, specifically, a U(1) gauge invariance. The U(1) refers to the fact that the phase transformation is unitary (does not change the normalization of the wave function) and that the integer Q that appears in it is just a number, a 1×1 matrix.

The gauge invariance is such a powerful symmetry that it, together with the requirement that the coupling constant be dimensionless, completely determines the form of the interaction. Because of the symmetry, the photon field, $\mathbf{A}(x)$, is coupled to each particle field with a strength eQ; the symmetry also ensures that the photon is massless.

Because of the central role played by gauge invariance in the electromagnetic interactions, the photon field $\mathbf{A}(x)$ is called a "gauge field" and the photon a "gauge particle."

Chromodynamics

The quark fields are three-component vectors in a "color space," the components corresponding to the colors red, white and blue. (We shall denote such "color vectors" by arrows: $\vec{q}$). The force between quarks is mediated by a set of "gluon fields" analogous to the photon field of electrodynamics) whose coupling strength is given by a set of color charges, T_a. These charges are 3×3 matrices (traceless and Hermitian) in color space. There are exactly eight independent such matrices, and there are eight different types of color charges T_a, and eight different types of gluon fields G_a. The eight matrices T_a are 3×3 analogs of the famous 2×2 Pauli matrices, σ_i; to normalize them one generally requires that the trace of T_a should be one-half. The color charges are also additive, and particles and antiparticles have opposite color charges.

When a quark is shaken it can emit any one of the eight gluons. The strength of the interaction (analogous to eQ) between a quark and the gluon field $\mathbf{G}_a$ is $g_3\mathsf{T}_a$. Two of the matrices T_a, conventionally called T_3 and T_8, are diagonal, so that quarks that emit gluons $\mathbf{G}_3$ or $\mathbf{G}_8$ do not change their color; the other gluon fields cause transitions between colors. (See Figure 1.)

Quantum chromodynamics also has a gauge invariance. We can add to the gluon field a set of gradients (and simultaneously perform unitary transformations on the matrices) and change the phases of the quark fields without changing the theory. Because the fields are matrices, the gauge transformations are more complex than they are in electrodynamics; for example, the quark fields transform with a matrix phase factor

$$q(x) \rightarrow \exp[ig_3\Sigma_a\theta_a(x)\mathsf{T}_a]\vec{q}(x).$$

Because only two of the matrices are diagonal, the colors of quarks can be mixed up differently at each point in space—and all the physical results of the theory remain the same.

The net effect of the virtual exchange of all these gluons is to produce a force that tends to bind quarks together into systems that are color-neutral, that is, for which $\vec{q}^{\,\dagger}\mathsf{T}_a\vec{q}$ is zero ($a = 1, \ldots, 8$). The simplest such systems are quark-antiquark pairs (the π^+, for example, consists of a u and a d), and quark triplets (the proton, for example, contains two

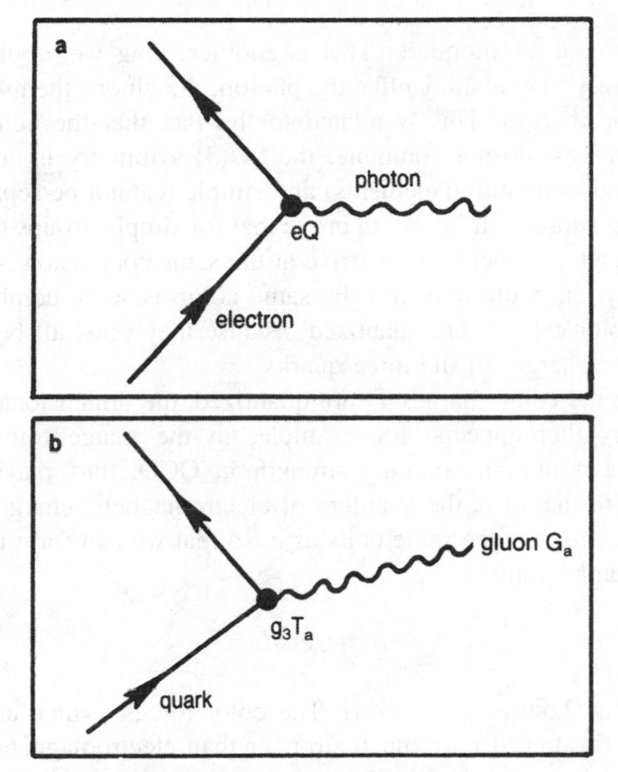

FIGURE 1. *Fundamental interactions (a) in QED and (b) in QCD. In (a) an electron emits (or absorbs) a photon, changing its energy and momentum; the interaction strength is given by* eQ *(Q is the dimensionless charge, an integer). In (b) a quark emits (or absorbs) a gluon, changing its energy, momentum, and possibly its "color."*

u's and a d, one quark of each color); the color-neutrality of quark triplets reflects the tracelessness of the T_a matrices.

Actually, we think that the binding of quarks and gluons into color-neutral systems is absolute: The quarks and gluons are permanently confined within hadrons. There is thus no hope of ever seeing an isolated quark or gluon.

In addition to the color charges, the quarks have electromagnetic charges, Q. The u quark for example has $Q = \frac{2}{3}$ and the d quark has $Q = -\frac{1}{3}$. Thus, the proton has total charge $Q = 1$ and the neutron (which is two d's and a u) has total charge $Q = 0$. Notice that, though the quarks have fractional electric charge, the fractional charges are just right to give integral charge to the color-neutral systems of three quarks or a quark and an antiquark. This is another thing we would like our unified theory to explain. Unlike the photon, the gluons themselves must carry color charges. This is related to the fact that the color charges (being matrices) do not commute; the SU(3) symmetry is non-Abelian (contains noncommuting elements) and simple (cannot be separated into commuting subsets). It is easy to prove that for simple groups the charges are quantized. Another way to arrive at the same conclusion is to realize that any system with color has the same color as some combination of quarks. Color charges are quantized because they must all be multiples of the color charges of the three quarks.

Because the color charges T_a are quantized, the dimensional coupling constant g_3 that appears, for example, in the gauge transformation measures a minimum coupling strength in QCD, thus playing a role analogous to that of e, the quantum of electromagnetic charge, in QED. The constant g_3 has the same units as e, so that we can form the dimensionless combination

$$\alpha_3 = g_3{}^2/\hbar c.$$

Experiments show that $\alpha_3 >> \alpha$: The color forces (which are "strong interactions" after all) are much stronger than electromagnetic forces.

Asymptotic Freedom

The relativistic quantum-mechanical vacuum is a lot more than just an empty space. It is seething with virtual particles. For example, the same electromagnetic interaction that allows an electron to emit a photon also allows an electron, a positron and a photon to appear out of nothing. Of course, these cannot be real particles because energy is not conserved, but they can exist as virtual particles for a short time consistent with the uncertainty principle.

If a real charged particle is added to this complicated vacuum, it can polarize the virtual electron-positron pairs. If the real particle has a positive charge, the virtual positive charges are pushed away from it slightly while the virtual negative charges are pulled toward it. The net result is that some of the virtual positive charge is pushed far away and the real charge is surrounded by a negatively charged vacuum.

Suppose now that we want to measure the charge on one of these particles. The standard way is to measure the electric field on a sphere of radius r around the charge and use Gauss's law to find the charge enclosed. But, because of the polarization of the vacuum, this charge will vary with r: It will decrease as r increases because the sphere contains more and more negatively charged vacuum.

You might think that we could find the "bare" particle charge by going in very close, making r very small. But that doesn't work. The density of negative charge in the vacuum increases as you get closer to the real charge, and no matter how small you make the sphere, there is always a significant amount of negative charge in the enclosed vacuum (this can only happen if the "bare" positive charge is infinite and the density of negative charge in the vacuum goes to infinity as r goes to zero).

This sounds dangerous, but it is okay so long as the charge in any finite sphere is finite. After all, in any experiment, the charges are separated by some characteristic distance d. For example, in an atomic physics experiment a typical distance is $d \sim 10^{-8}$ cm. The appropriate "charge" to use for describing the electromagnetic interactions in such an experiment is the charge inside a sphere of radius d around the point charges.

This process of replacing the infinite bare charge with a finite charge measured at a given distance from the point charge is called "renormalization;" it works only for a specific class of field theories. We shall deal only with such "renormalizeable" theories.

The renormalization does not affect charge quantization: The charge in a sphere of radius d around an electron is the same as the charge in a sphere of radius d around the muon (there are some effects due to the mass difference between μ and e, and the proton is a bit more complicated because of the strong interactions, but never mind). This means that the dimensionless charge Q still takes on integral values (except for quarks), but the coupling constant α now depends on d; its value is $1/137$ at an atomic distance, $\sim 10^{-8}$ cm.

In QED, α is small, so the dependence of α on d is weak—usually negligible, in fact—but it has been measured. (It contributes to the Lamb shift, for example.) In QCD, where α_3 is much larger, this effect is very important: A color-charged particle polarizes the vacuum to a substantial degree. Furthermore, the gauge particles that mediate the color force themselves carry color charge; the cloud of virtual gluons that surrounds a color-charged particle thus effectively spreads out the charge of the particle. It turns out that this effect overwhelms the effect of vacuum polarization, so that a positive color charge is surrounded by a positive charged vacuum (Figure 2). Because the charge is spread out, the color charge within a sphere of radius r decreases (very slowly) to zero as r goes to zero. The color forces get weaker at short distances! This property of chromodynamics is called asymptotic freedom. It was discovered by H. David Politzer at Harvard (now at the California Institute of Technology) and by David Gross and Frank Wilczek at Princeton. Asymptotic freedom has been tested successfully in recent experiments that probe very short distances.

Weak Interactions

The weak interactions are also described by a gauge theory. In this case the fundamental particles appear in doublets, so the invariance is an SU(2) symmetry. (This is the same symmetry group as the angular momentum of quantum mechanics.) The weak charges are the three independent 2×2 traceless, Hermitian matrices R_i (proportional to the Pauli matrices). Coupled to these three charges are three gauge particles, W (these are often called "intermediate vector bosons"). Again, because the charges are matrices, the emission or absorption of a W_i can change the identity of a particle. Thus, for example, an electron can emit a W_1 and turn into a neutrino.

The W_3 is electrically neutral, but the W_1 and W_2 carry electric charge. The fields corresponding to particles of definite charge are the combintions $W^{\pm} = W_1 \pm iW_2$.

The weak interactions have the strange property that they discriminate between particles of different handedness (that is, with spins directed along or against the direction of motion).

The handedness of a massive particle can of course be changed by bringing it to rest and accelerating it in the opposite direction without changing its spin. Thus, massive particles have both left-handed and right-handed components. However, the neutrinos apparently have zero rest mass; so like photons they always travel at the speed of light. Because a neutrino cannot be stopped, its handedness never changes. In fact, experimentally only left-handed neutrinos and right-handed antineutrinos have been observed. (The antiparticle of a right-handed particle is left handed.) Their oppositely spinning counterparts are presumed not to exist.

Some of the quarks and leptons carry no weak charge at all (they are "singlets" under the SU(2) symmetry) while the same particles with

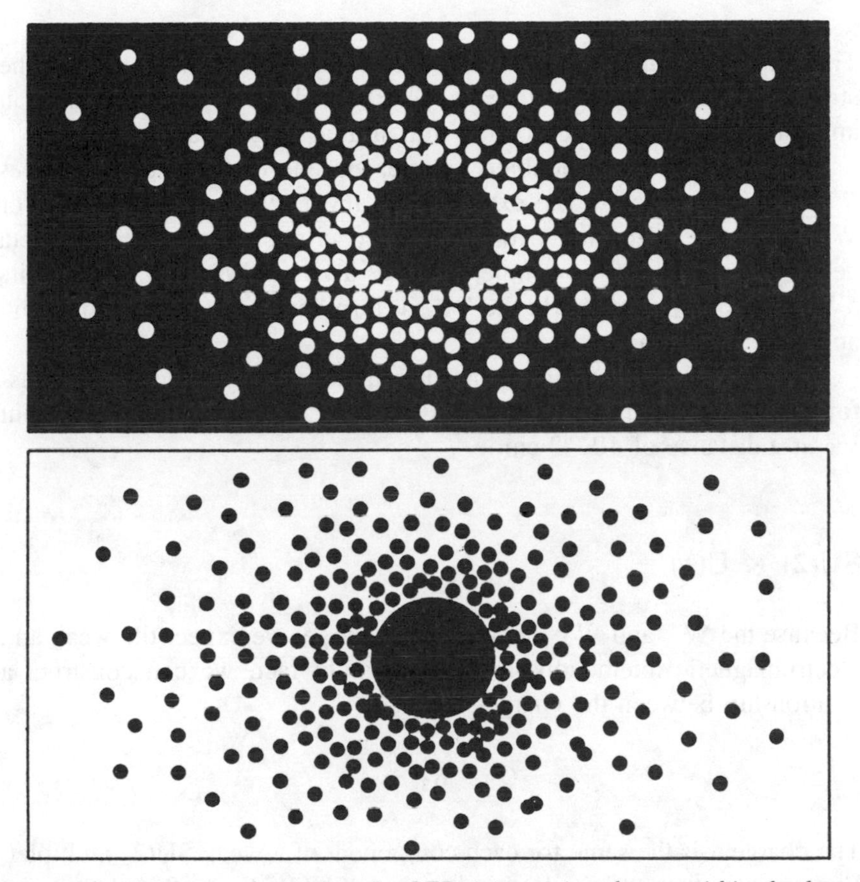

FIGURE 2. *Vacuum polarization. In QED (a) a point charge within the large spot is surrounded by a cloud of virtual charges of the opposite sign. In QCD (b) a point ''charge'' within the large spot is surrounded by virtual quarks and gluons with predominantly the same color as the central quark.*

oppositely directed spins are members of weak-charged doublets. The singlets include

$$e^-_R \quad e^+_L \quad u_R \quad \bar{u}_L \quad d_R \quad \bar{d}_L$$

and the corresponding doublets are

$$\begin{pmatrix} \nu_L \\ e^-_L \end{pmatrix} \quad \begin{pmatrix} e^+_R \\ \bar{\nu}_R \end{pmatrix} \quad \begin{pmatrix} u_L \\ d_L \end{pmatrix} \quad \begin{pmatrix} \bar{d}_R \\ \bar{u}_R \end{pmatrix}.$$

The subscripts indicate the left or right handedness of the particles. The strange, charmed, truth and beauty quarks and the leptons mu and tau and their associated neutrinos fall into similar singlets and doublet.

The weakly charged doublets couple, via one of the charge-matrices R_i, to the W particles. The exchange of a W^+ or W^- (or, of a W_1 or W_2) gives rise to the weak force observed in nuclear physics. The beta decay of a neutron, for example, involves a virtual W^-: a d quark emits a W^-, becoming a u quark; the W^- in turn decays into an electron and an antineutrino (Figure 3).

The dimensional coupling constant g_2 that appears in the gauge transformation has the dimensionless eqivalent $\alpha_2 = g_2^2/\hbar c$, which is about 1/30 at a distance of 10^{-16} cm.

SU(2) × U(1)

Because the W^+ and W^- carry electric charge, we expect the weak and electromagnetic interactions to be related. In fact, we can construct a relationship between the charges:

$$Q = R_3 + S.$$

The charge S is the same for each component of a weak SU(2) multiplet. For particles that are singlets under weak SU(2), $R_3 = 0$ and S is just the electric charge. For SU(2) doublets, $R_3 = \pm 1/2$ and S is the average electric charge of the doublet. For example, the right-handed electron has $R_3 = 0$, $S = -1$ and $Q = -1$; the left-handed electron neutrino and electron have $S = -1/2$ and, respectively, $R_3 = +1/2$ and $-1/2$ and $Q = 0$ and -1.

As for the other charges we have discussed, S is associated with a gauge invariance. Particles for which $S = 0$ can emit a gauge particle, which we shall call V; we shall denote the dimensionless coupling constant for this process α_1. But because the charges Q, R_3 and S are related, the three gauge fields (or, equivalently, the particles γ, W_3 and V) are related as well. The photon field (the vector potential, **A**) is a linear combination of the fields $\mathbf{W}_3$ and **V**:

$$\mathbf{A} = \mathbf{V} \cos \theta_w + \mathbf{W}_3 \sin \theta_w$$

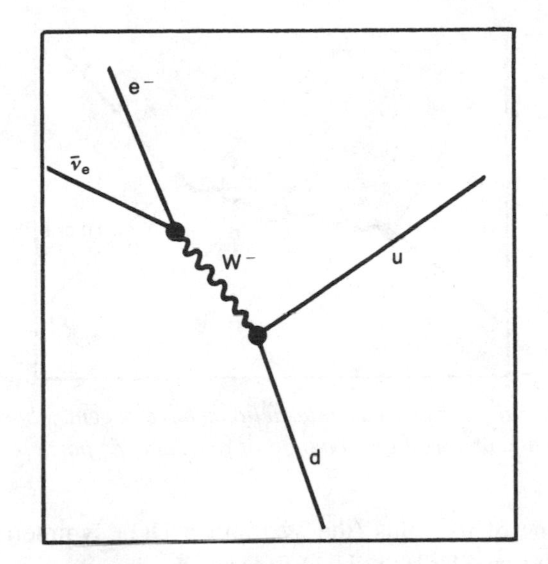

FIGURE 3. *Beta decay. A d quark within a neutron emits a virtual* W^-, *becoming a u (the neutron thus turns into a proton). The virtual* W^- *in turn decays into an electron and an antineutrino.*

There is, of course, another, independent linear combination of the V and W_3 fields; it is called the Z^0

$$\mathbf{Z}^0 = \mathbf{W}_3 \cos \theta_w - \mathbf{V} \sin \theta_w$$

The parameter θ_w is called the weak mixing angle; its value is determined by the coupling strengths e, g_2 and g_1.

Like the photon, the Z^0 has zero electric charge and can be emitted without changing a particle's identity. But, unlike the photon, the Z^0 can be emitted and absorbed directly by neutrinos. The virtual exchange of Z^0 gives rise to a weak force between neutrinos and the other particles; this is called a "neutral-current" interaction because the exchanged particle is electrically neutral (Figure 4). In contrast, the classic weak interaction shown in the graph above is called a "charged-current" interaction. The neutral-current interaction is very different and much harder to observe because no change in particle identity is involved. Recent experiments, however, have in fact seen the results of weak neutral currents, thus confirming this aspect of the unified theory of weak and electromagnetic interactions. Because the theory combines an SU(2) sym-

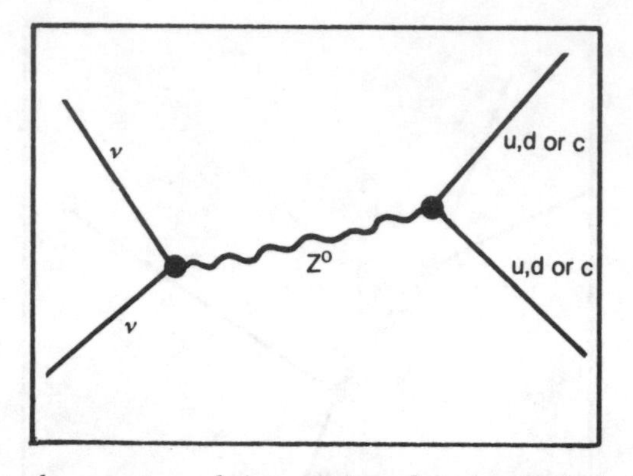

FIGURE 4. *Neutral-current weak interactions have recently been detected experimentally. They involve the exchange of a virtual Z^0 particle.*

metry for some of its fields (the W_i) and a U(1) symmetry for another (V), it is called an SU(2) × U(1) theory.

Spontaneous Symmetry Breakdown

Something crucial is missing from the above description of weak and electromagnetic interactions. The electromagnetic force has an infinite range ($F \propto 1/r^2$), and is mediated by a particle (the γ) with zero mass. The weak interactions, however, have a very short range ($\sim 10^{-16}$ cm), and the exchanged particles, the W^+, W^- and Z^0 must be massive—indeed about 100 times more massive than a proton. What has happened to gauge invariance, which tells us that the gauge particles have zero mass? The answer is that the underlying force law is symmetrical but the vacuum is not. The vacuum distinguishes the W^+, W^- and Z^0 from the photon, making them heavy but leaving the photon massless. The structure of the vacuum breaks the SU(2) × U(1) gauge invariance and allows the gauge particles to develop mass. Such a situation is called a spontaneous symmetry breakdown.

A useful analogy is the breakdown of rotational invariance in a crystal. Consider, for example, a grain of salt. It is built up out of sodium ions and chloride ions. The forces between the ions are electromagnetic forces, which do not pick out any special directions in space. However, when the ions are packed together, they form a cube. If you lived inside a grain of salt, you would find that your space does have some special directions; the directions perpendicular to the faces of the cube, for example, have special properties. It is true that a "giant" living outside the cube of salt could pick it up and rotate it (and you with it) without affecting the physics of your world, but to you inside the cube, the rotational symmetry of the laws of physics would not be obvious. The rotational symmetry has been spontaneously broken by your environment, the salt crystal.

The relativistic quantum-mechanical vacuum, in which we all live, is like the salt crystal. The symmetry which is broken spontaneously is not rotational invariance, but the gauge invariance which, if unbroken, would make the W^+, W^-, and W^0 interchangeable and massless. The vacuum picks out a direction in this "gauge space" so that the W^0 direction is singled out. In the process the W^+, W^-, and the Z^0 (a particular combination of W^0 and V^0, remember) get a mass, while the photon remains massless because the electromagnetic gauge invariance is not broken by the vacuum.

The gauge theory of the weak and electromagnetic interactions, with spontaneously broken $SU(2) \times U(1)$ symmetry, was worked out in the 1960s by Glashow at Copenhagen (now Harvard), Steven Weinberg at MIT (now also at Harvard) and Abdus Salam at Imperial College, London, and the International Centre for Theoretical Physics in Trieste. Glashow worked out the form of the gauge theory but did not know how to give mass to the W and the Z. Weinberg and Salam worked out the effect of the spontaneous symmetry breakdown (using a mechanism developed earlier by Peter Higgs at the University of Edinburgh and T.W.B. Kibble at Imperial College) and produced a consistent theory. Glashow, Salam and Weinberg received the Nobel Prize for this work in 1979.

There is another aspect of the spontaneous breakdown of the gauge symmetry which can be understood in the crystal analog. It is "graininess." The salt crystal, after all, is grainy on an atomic scale 10^{-8} cm. Living inside the crystal, you might well distinguish three different domains of distance, associated with different physics:

- Distances much smaller than 10^{-8} cm. If you do an experiment that probes the structure of your world at distances much smaller than 10^{-8} cm, it doesn't matter very much that you are living in a crystal. You will find that the results of your experiment will be approximately rotationally invariant. This is because the nonuniform electric fields that mess up the crystal's rotation invariance are very weak compared with the kinds of fields you have to generate to probe subatomic distances.

Once you are inside the atom, it is the constituents of the atom that matter, not how the atom is put together with other atoms in the crystal.

- Distances of the order of 10^{-8} cm. Here you will see all of the complicated interatomic forces that are responsible for packing the atoms into a crystal.
- Distances much larger than 10^{-8} cm. Here you will see the cubic structure of the crystal; there is no apparent rotational invariance that one can see at this scale. All that remains is the discrete symmetry group associated with the crystal structure.

Just as a crystal is grainy at an atomic scale, the vacuum is grainy at a smaller scale, on the order of 10^{-16} cm, associated with the spontaneous breakdown of the SU(2) × U(1) symmetry. Again, we can identify three distinct regions of distance scale:

- Distances much smaller than the 10^{-16} cm. Here you will see the SU(2) × U(1) gauge structure of the world as an explicit (though approximate) symmetry. At such short distances, the masses of the W^+, W^-, and Z^0 (which are around 100 GeV) are negligible compared to the energies required to do the experiment. (To probe down to $\Delta x \sim 10^{-18}$ cm requires an energy about 10^4 GeV.) Thus, the fact that the W and Z^0 are massive while the photon is not can be ignored. All the particles are light compared to the energy of the experiment.
- Distances of the order of 10^{-16} cm. Here your life will be very complicated. You will see the physics responsible for spontaneously breaking the SU(2) × U(1) symmetry. W's and Z^0's will show up in your experiments, but they will look very different from photons, because their rest energies are not small compared to the energies of your probes.
- Distances much greater than 10^{-16} cm. Here you will not see the SU(2) × U(1) gauge symmetry at all. You will not even see W's and Z^0's directly. You do not have a probe with high enough energy to produce them. You see the unbroken electromagnetic gauge invariance directly. But the heavier gauge particles only show up in the weak short-range interactions caused by their virtual exchange. These interactions would hardly be noticeable except that they do things the electromagnetic and strong interactions do not do. The W^+ and W^- exchanges change particle identities. These show up in beta decay. The Z^0 exchanges cause neutral-current interactions of neutrinos. And they all violate parity sym-

metry because the left- and right-handed components of the quarks and leptons interact differently.

This last region is the domain of contemporary particle-physics experiments; indeed, all of the short-range weak effects we have described have actually been observed. The fact that everything fits together as expected gives us confidence that the picture of weak interactions as a spontaneously broken SU(2) × U(1) gauge theory is correct, even though no one has seen a W or a Z^0.

A Unified Theory

The SU(2) × U(1) gauge theory we have just described is a partial unification of the weak and electromagnetic interactions. It describes the charged-current and the neutral-current weak interactions and the electromagnetic interactions, which at first sight look very different as gauge interactions. But, although the theory is unified, there are still two different interactions involved, associated with the SU(2) and the U(1) groups. There are two dimensionless coupling constants: α_2 for the SU(2) gauge particles (the three W's) and α_1 for the U(1) gauge particles (the V^0). Their values can be determined experimentally because α, the ordinary electromagnetic coupling constant, is given by $1/\alpha = 1/\alpha_1 + 1/\alpha_2$ and the weak mixing angle θ_w is given by $\sin^2\theta_w = \alpha/\alpha_2$.

Also, the electric charge is only partially quantized. Within an SU(2) multiplet (such as the doublet e_L^-, ν_L) all charge differences are integers, because a particle in a multiplet can emit a $W^\pm$ and become another member of the multiplet. But the average charge of the multiplet is not quantized. The average charge is the U(1) charge S, and, as in the case of the electromagnetic charge in the QED theory, there is no theoretical reason for it to be quantized. And finally, there is still no connection between quarks and leptons. But, now we have all the pieces to the puzzle, and we can fit them together into a unified theory.

If we merely combined the three gauge theories we would have a theory whose symmetry is SU(3) × SU(2) × U(1). Such a combination of symmetries can be components of an SU(5) symmetry. This is a symmetry group whose fundamental representation has five elements (like the triplets of colored quarks or the weak doublets), and whose charges are 5 × 5 traceless Hermitian matrices.

The five-element vectors that form the basis of the group representation (analogous to the color triplets of quarks, for example) contain both quarks and leptons, but only of a single handedness. One of these, for instance, is

$$d_R^r \quad d_R^w \quad d_R^b \quad e_R^+ \quad \overline{\nu}_R$$

All of these are related by an SU(5) symmetry; the first three are related by the SU(3) color symmetry, the last two by the weak SU(2).

The 24 SU(5) matrices are coupled to 24 gauge fields. Eight of the matrices have elements only in the upper left 3 × 3 submatrix; these are coupled to the eight gluons. Four of the matrices are coupled to the $W^{\pm}$, γ and Z^0. The photon, for example, couples to

$$\boldsymbol{Q} = \begin{pmatrix} -1/3 & 0 & 0 & 0 & 0 \\ 0 & -1/3 & 0 & 0 & 0 \\ 0 & 0 & -1/3 & 0 & 0 \\ 0 & 0 & 0 & 1 & 0 \\ 0 & 0 & 0 & 0 & 0 \end{pmatrix}.$$

The W particles couple to charges that contain nonzero elements only in the lower right 2 × 2 submatrix.

The remaining 12 gauge particles cause transitions between the quarks and antileptons of the fundamental five. We shall call them X's.

Because SU(5) is a simple group, all the charges—including, of course, $\boldsymbol{Q}$—are quantized. We have thus finally arrived at a theoretical understanding of charge quantization.

At least, we have arrived at a reason for charge quantization of the right-handed d quarks, positrons and antineutrinos—and, consequently, of their left-handed antiparticles. But something even better is in store for us. The next simplest family with an SU(5) gauge invariance can be built out of the five fundamental objects by forming 10 pairs. (In an analogous way, a spin-zero state can be formed from a pair of spin-½ states.) The 10 states formed from pairs of right-handed states have all the right charges (color, electric and weak) to form the remaining left handed particles and antiparticles; those formed from pairs of left-handed states make up the remaining right-handed states. The process is illustrated in Table 1. To clarify the way individual, states are built up out of the fundamental 5 (and their antiparticles, called $\overline{5}$) we have denoted

TABLE 1. The Fundamental Particles in SU(5)

Particle	*Representation*	*"Metacolor"*	Q	T_3	T_8	R_3
d_R^r	5	a	$-1/3$	$1/2$	$1/2\sqrt{3}$	0
d_R^w	5	b	$-1/3$	$-1/2$	$1/2\sqrt{3}$	0
d_R^b	5	c	$-1/3$	0	$-1/\sqrt{3}$	0
e_R^+	5	d	1	0	0	$1/2$
ν_R	5	e	0	0	0	$-1/2$
u_L^r	10	a + d	$2/3$	$1/2$	$1/2\sqrt{3}$	$1/2$
u_L^w	10	b + d	$2/3$	$-1/2$	$1/2\sqrt{3}$	$1/2$
u_L^b	10	c + d	$2/3$	0	$-1/\sqrt{3}$	$1/2$
d_L^r	10	a + e	$-1/3$	$1/2$	$1/2\sqrt{3}$	$-1/2$
d_L^w	10	b + e	$-1/3$	$-1/2$	$1/2\sqrt{3}$	$-1/2$
d_L^b	10	c + e	$-1/3$	0	$-1/\sqrt{3}$	$-1/2$
e_L^+	10	d + e	1	0	0	0
$\bar{u}_L^r$	10	b + c	$-2/3$	$-1/2$	$-1/2\sqrt{3}$	0
$\bar{u}_L^w$	10	a + c	$-2/3$	$1/2$	$-1/2\sqrt{3}$	0
$\bar{u}_L^b$	10	a + b	$-2/3$	0	$1/\sqrt{3}$	0
$\bar{d}_L^r$	$\bar{5}$	$\bar{a}$	$1/3$	$-1/2$	$-1/2\sqrt{3}$	0
$\bar{d}_L^w$	$\bar{5}$	$\bar{b}$	$1/3$	$1/2$	$-1/2\sqrt{3}$	0
$\bar{d}_L^b$	$\bar{5}$	$\bar{c}$	$1/3$	0	$1/\sqrt{3}$	0
e_L^-	$\bar{5}$	$\bar{d}$	-1	0	0	$-1/2$
ν_L	$\bar{5}$	$\bar{e}$	0	0	0	$1/2$
$\bar{u}_R^r$	$\overline{10}$	$\bar{a} + \bar{d}$	$-2/3$	$-1/2$	$-1/2\sqrt{3}$	$-1/2$
$\bar{u}_R^w$	$\overline{10}$	$\bar{b} + \bar{d}$	$-2/3$	$1/2$	$-1/2\sqrt{3}$	$-1/2$
$\bar{u}_R^b$	$\overline{10}$	$\bar{c} + \bar{d}$	$-2/3$	0	$1/\sqrt{3}$	$-1/2$
$\bar{d}_R^r$	$\overline{10}$	$\bar{a} + \bar{e}$	$1/3$	$-1/2$	$1/2\sqrt{3}$	$1/2$
$\bar{d}_R^w$	$\overline{10}$	$\bar{b} + \bar{e}$	$1/3$	$1/2$	$1/2\sqrt{3}$	$1/2$
$\bar{d}_R^b$	$\overline{10}$	$\bar{c} + \bar{e}$	$1/3$	0	$1/\sqrt{3}$	$1/2$
e_R^-	$\overline{10}$	$\bar{d} + \bar{e}$	-1	0	0	0
u_R^r	$\overline{10}$	$\bar{b} + \bar{c}$	$2/3$	$1/2$	$1/2\sqrt{3}$	0
u_R^w	$\overline{10}$	$\bar{a} + \bar{c}$	$2/3$	$-1/2$	$1/2\sqrt{3}$	0
u_R^b	$\overline{10}$	$\bar{a} + \bar{b}$	$2/3$	0	$1/\sqrt{3}$	0

the analogs of the SU(3) "colors" (r, w, b)—the "metacolors," so to say—by letters a–e. The table also gives the values of the (diagonal) color charges T_3 and T_8 and of the (diagonal) weak charge R_3 for each of the particles. (This is, of course, not in any sense a dynamical building process.)

In the language of group theory, the quintuplets 5 and $\overline{5}$ are the smallest irreducible representations of SU(5). The antisymmetric combination of two 5's (or two $\overline{5}$'s) is a 10 (or a $\overline{10}$) and is the next-smallest representation of SU(5). These representations are all we need to describe all the quarks and leptons. The other leptons and quarks with other flavors also have the same SU(5) representations: The quarks s and c together with the leptons μ and ν_μ form another SU(5) "family," as do the quarks t and b with the τ and ν_τ. There is, in fact some astrophysical evidence that there are not very many such families.

It is important to realize that this building process did not have to work. It represents the first, the simplest, and in some ways the most remarkable success of the SU(5) unification.

Providing a reason for charge quantization is, to us, one of the nicest aspects of the theory. Charge quantization follows simply from the building process. Even if more complicated SU(5) families exist, Q must still be quantized in multiples of ⅓, because all charges must be sums of the charges of the simplest family. Furthermore, the fact that all observed systems have integral Q is connected with their color neutrality. The only way to "build" color-neutral systems out of the simplest family is to use the particles e_R^+ or $\overline{\nu}_R$ or the combination $d_R^r + d_R^w + d_R^b$, each of which has integral charge.

In SU(5) charge quantization is trivial because the charges in the 5 are "commensurate." Actually, however, the quantization of charge is somewhat deeper: It has a topological aspect and is related to the existence of magnetic monopoles in gauge theories that are based on simple groups, which in turn requires charge quantization.

This SU(5) unification was worked out by us at the end of 1973.

The Unification Scale

All of the charges in SU(5) are treated symmetrically, so in some sense, there ought to be only one coupling constant describing all of the interactions. But that is certainly not the situation at distances of the order of 10^{-16} cm and larger. There α_3 is greater than α_2 which, in turn, is greater than α_1. For this reason, and for others to which we will come below, there must be another level of spontaneous symmetry breakdown, this one associated with the breaking of SU(5) down to SU(3) × SU(2) × U(1).

We must assume that the vacuum is grainy and structured at another scale L much smaller than the 10^{-16} cm associated with spontaneous breakdown of SU(2) × U(1). As before, there are three different regions of distance scale:

• Distances much less than L. Here you will see the SU(5) gauge invariance as an explicit, approximate symmetry. All the gauge particles, the photon, the gluons, the W's, and Z° and the X's are light compared to the energies required to probe these short distances.
• Distances of the order of L. At these distances you will see the complicated physics of the spontaneous breakdown of SU(5). The X's will be produced, but will be much heavier than all the other gauge particles, with a mass of order $\hbar c/L$.
• Distances much greater than L. Here you will see only the SU(3) and the SU(2) × U(1) gauge invariance directly. The X's are too heavy to be produced directly, but they do give rise to very weak, very short-range interactions.

We can now understand the disparity in coupling constants. In the first region, at very short distances where the SU(5) gauge symmetry is explicit, there is indeed only a single coupling constant describing all of the gauge interactions. But in the second region the symmetry is broken so that for larger distances the couplings diverge. Since the SU(3) coupling constant α_3 is the most asymptotically free, it increases faster than α_2 as the distance at which it is measured increases. The U(1) coupling constant α_1, on the other hand, is not asymptotically free at all (because the gauge particle does not carry the charge). So, α_1 actually decreases as the distance increases. This situation is depicted in Figure 5.

Figure 5 shows, qualitatively, that at distances much larger than L, the couplings have the right form, with α_3 greater than α_2 greater than α_1. But we can do better. At distances smaller than 10^{-16} cm, all of the couplings are rather small, and their dependence on the distance at which they are measured is actually calculable. Then, if two of the couplings at 10^{-16} cm are known, they can be followed to shorter and shorter distances until they meet. This will be a distance of the order of L. Furthermore, one can then predict the values for distances larger than L. For example, one can use α (the combination of α_2 and α_1 that occurs in the electromagnetic interactions), which is known very well experi-

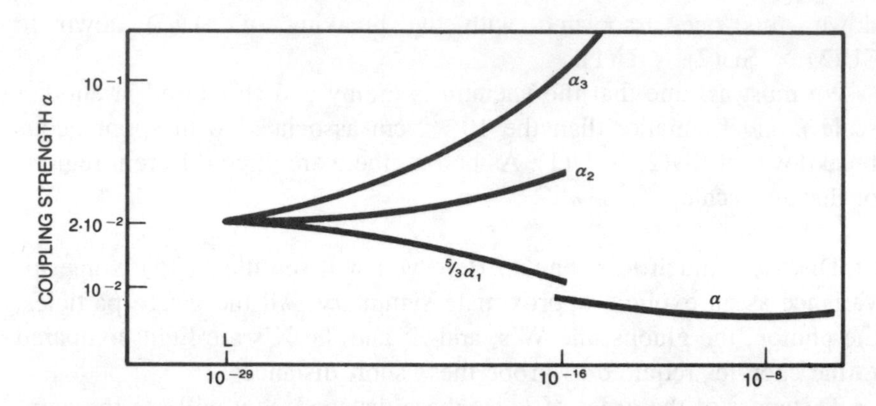

FIGURE 5. *Coupling strengths vary with distance. At distances up to the "unification distance" L all three constants are the same (up to a group-theoretic factor of 5/3 for α_1). The strong coupling, α_3, becomes large at distances greater than 10^{-14} cm, signalling the breakdown of QCD perturbation theory due to quark confinement. At distances larger than 10^{-16} cm the spontaneous breakdown of SU(2) × U(1) collapses α_2 and α_1 into the single electromagnetic coupling α.*

mentally, and α_3 (the gluon coupling constant), which has been measured, but with rather large errors, and then estimate L and θ_w. The results are that L is about 10^{-29} cm and $\sin^2\theta_w$ is about 0.20.

This prediction was worked out at Harvard in 1974 by Georgi, Helen Quinn (now at SLAC) and Weinberg for a large class of unified theories, including SU(5). At the time, it looked bad for unification because the experimental value of $\sin^2\theta_w$ was about 0.35.

This discrepancy between the predicted and observed values didn't bother the two of us at the time because we were not certain that the SU(3) and the SU(2) × U(1) were correct descriptions of physics at 10^{-16} cm, so we thought that SU(5) might have to be generalized to include new physics. However, in the years between 1974 and the present, the SU(3) and SU(2) × U(1) theories have passed a number of important experimental tests so that today we are much more confident in the underpinnings of the SU(5) theory. And, fortunately for the theory, the experimental value of $\sin^2\theta_w$ has moved down, and today is 0.23 ± .02, almost in agreement with the unification prediction.

Proton Decay

The X particles in the SU(5) unified theory are very heavy because they are associated with a spontaneous symmetry breakdown at a very short distance. Their mass is about $\hbar c/L$, or 10^{15} times the mass of the proton. We can never produce them in accelerators, but it may be possible to observe the effect of their virtual exchange. The interactions that are mediated by X exchange are very weak because they have such a short range. But, like the ordinary weak interactions, these very weak interactions cause processes that could not occur at all without them.

The most interesting of these are interactions that change baryon number. Baryon number is defined as one-third the number of quarks minus one-third the number of antiquarks. The baryon number of a proton is one while the baryon number of a meson or a lepton is zero.

The SU(3) and the SU(2) × U(1) gauge interactions do not change baryon number: They conserve it, because they gauge particles do not cause transitions between quark and lepton or quark and antiquark. If baryon number were exactly conserved, the proton would have to be absolutely stable because it is the lightest particle with baryon number one. Because proton decays have not been observed (although some experimenters have tried to find them) it was quite reasonable to assume that baryon number is absolutely conserved, and most physicists did.

The processes that change baryon number, which are caused by the X particles in the SU(5) theory, can, however, lead to proton decays. One of the X particles, for example is coupled to "metacolors" *a* and *d*. A d^r_R quark can emit such an X change into an e^+; also, a u^r_L quark can absorb the same X and change into a $\bar{u}^b_L$. Exchanging an X can thus turn a pair of quarks, d and u, into a lepton and an antiquark, e^+ and $\bar{u}$. The baryon number of the system changes by one unit, from ⅔ to −⅓.

Figure 6 shows what happens if this process takes place inside a proton: It decays into a positron and a neutral pion. The decay is extremely rare. Because the X is so heavy, two quarks must come very close together to exchange it.

Quinn, Weinberg, and Georgi used the calculation of L to estimate the proton decay rate. This estimate has since been refined by many people, including Andrzej Buras, John Ellis, Mary Gailliard, and Demetres Nanopoulos, from CERN, Terry Goldman and Douglas Ross from the California Institute of Technology and William Marciano from Rockefeller. The present estimate is the that the decay rate is about one decay per proton every 10^{31} years.

Salam and Jogesh Pati of the University of Maryland independently, and in a different context, suggested that the proton might be unstable. Their model is based on a different version of the QCD theory of hadrons.

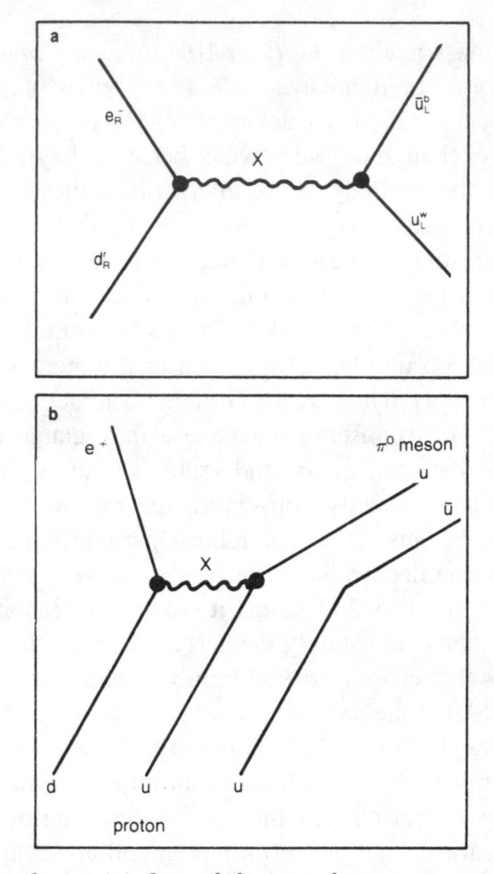

FIGURE 6. *Proton decay. (a) One of the superheavy gauge particles X can be exchanged when two quarks come within about 10^{-29} cm of each other. (b) When such an exchange occurs in a proton it decays into a positron and a pion.*

What is the chance of observing proton decay directly? How is it possible to measure a proton lifetime of about 10^{31} years when the age of the universe since the Big Bang is only about 10^{10} years? The answer is that you get yourself a lot of protons. For example, in 1,000 tons of matter there are about 5×10^{32} protons and neutrons. We expect about 50 of these to decay each year. So you simply have to monitor everything that goes on inside 1,000 tons of matter and distinguish proton and neutron decays from everything else, and you should detect baryon number changing processes.

Several groups are planning experiments on this scale. The experiments are to be done underground to minimize the confusion caused by cosmic rays interacting in the sample of matter. One experiment is planned to take place in a salt mine near Cleveland, another in a silver mine in Utah, and yet another in an iron mine in Minnesota. On a slightly smaller scale are European experiments planned for tunnels under the Alps, an ongoing experiment in a gold mine in South Dakota, and an Indian-Japanese collaboration in the Kolar gold field.

Another very exciting idea has come out of the unification: Although we cannot build such a machine ourselves, there is one machine that may have directly probed the region of distances on the order of 10^{-29}

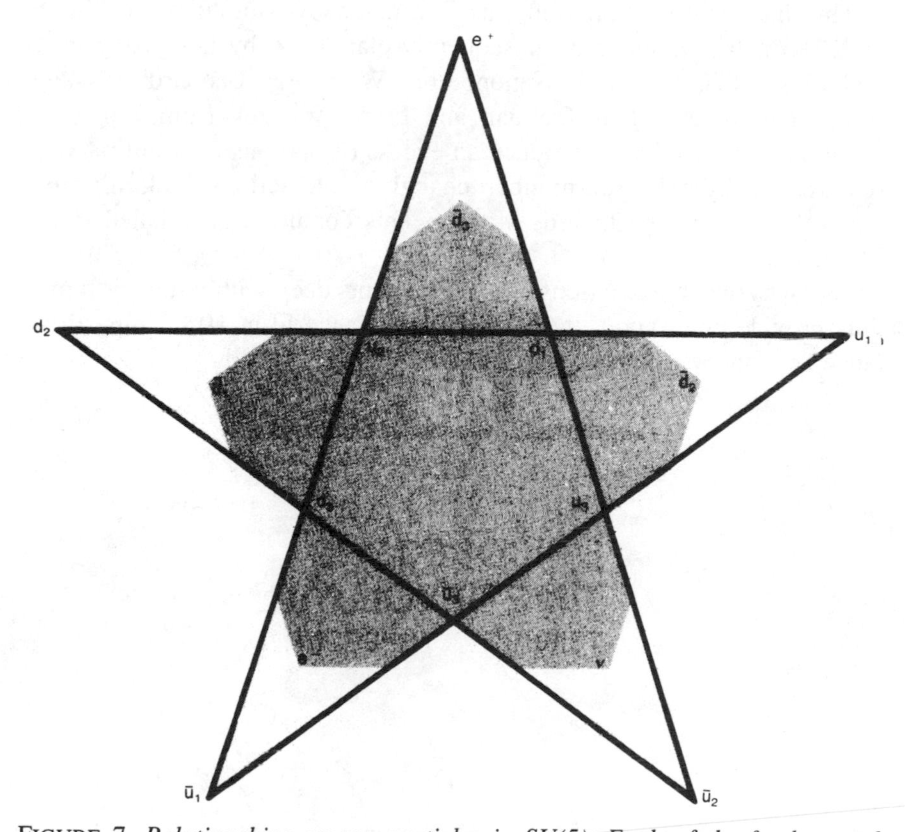

FIGURE 7. *Relationships among particles in SU(5). Each of the fundamental particles has some combination of the four independent charges and can be plotted as a point in a four-dimensional space. The diagram shows a two-dimensional projection of such a plot which preserves some of the symmetry of the four-dimensional original. The states shown are all left-handed particles. The pentagon represents the group of states that transforms according to the* "$\bar{5}$" *representation of SU(5), while the five-pointed star and the inner pentagon represents the* "10." *The vertical axis is a linear combination of* T_8, R_3 *and* Q; *the horizontal axis is a combination of* T_3 *and* R_3.

cm—the Big Bang of the universe itself. About 10^{-40} sec after the singularity that signaled the beginning of things the universe had expanded to a radius comparable to *L*. In this unimaginable era, the breakdown of the unifying gauge invariance was just beginning to appear, and the interactions that change baryon number were as strong as anything else. It seems possible that they may have produced more baryons than antibaryons. If so, unification is the solution to an old puzzle in astrophysics: Why is the universe built out of matter rather than antimatter?

This interesting speculation was first made by Motohiko Yoshimura of KEK in Japan, and was subsequently elaborated by many physicists including Ellis, Gailliard, Nanopoulos, Weinberg, Leonard Susskind from Stanford and Sam Treiman and Frank Wilczek from Princeton. They have shown that to produce an excess of baryons over antibaryons requires that the baryon-number changing interactions look different when they are run backwards in time. This condition is satisfied in the SU(5) theory.

Astrophysical observations and experiments deep within the earth may thus provide clues about the structure of the world at 10^{-29} cm, a distance so small we could never hope to probe it directly.

Grand Unification

Arthur, an intelligent alien from a distant planet, lands on earth and spies two humans playing chess. He sets himself two problems: to determine the rules of the game, and to find a winning strategy. My metaphor describes the two great divisions of physical science. In disciplines such as atomic physics, solid-state physics, or chemistry, the rules are known perfectly well; the building blocks are electrons and atomic nuclei, the forces are electromagnetic, and the laws are quantum mechanics. The problem is ''merely'' to apply the laws to the complex issues at hand. Nonetheless, new and exciting discoveries remain to be made; it is not enough to know the rules of chess to be a grand master.

In elementary-particle physics and cosmology, things are quite different. We don't know all the rules, nor even if they are knowable. The subject of ''grand unification'' is at this speculative frontier. It is an ambitious attempt to find fundamental unity among the diverse strands of elementary-particle physics. It may be a step in the right direction, or it may be quite wrong, but the theory will soon be put to a decisive experimental test. For me, these are very exciting times, and I would like to share the excitement with you.

Most physicists have an unshakeable faith in the underlying simplicity of nature; it is one of our most powerful guiding principles. Time and time again, this blind faith has been vindicated. Many of nature's bewildering tricks have been explained. What could be more different than magnetism, electricity, and light? Yet, in the nineteenth century, James Clerk Maxwell showed that these phenomena were simply different manifestations of the same fundamental laws. He described all these, as well as radio waves, radar, and radiant heat, by a unique and elegant system of equations. Maxwell's electromagnetism is the operative force controlling everything that we see, hear, smell, feel or taste.

Albert Einstein searched in vain for an ultimate simplicity wherein all the forces of nature could be treated by a single theory. He tried to unify electromagnetism with gravitation. We are still looking for such a unified theory, but we now know some of the reasons for Einstein's failure. There are simply more forces in heaven and earth than were dreamt of in Einstein's philosophy. He was loath to consider what were once the impenetrable mysteries of the atomic nucleus. "These questions would come out in the wash," he may have felt, "if only gravitation and electromagnetism could be put together."

The atomic nucleus is no longer such a mystery, but we have learnt that two additional forces are needed to explain its behavior. A correct unified theory must include a description of the strong and weak nuclear forces, which are responsible for holding the nuclear constituents together and which permit the radioactive transmutation of one chemical element into another. Einstein made another critical omission. He never completely accepted the quantum theory for which he was partly responsible. "God does not play dice," said Einstein. But we are now convinced that He does.

Today's approach to the construction of a unified theory depends on two fundamental constructs: Einstein's special theory of relativity and quantum mechanics. These are the two great revolutionary pillars of early twentieth-century physics. No development in physics in the past half century can be compared with these great accomplishments. The synthesis of these disciplines is called "quantum field theory," and it is within this esoteric context that all of elementary-particle physics is expressed.

The first successful quantum field theory is called quantum electrodynamics, or QED, and describes the interaction between electrons and photons (or particles of light). It is not a complete theory of elementary-particle physics, as it does not describe the structure of the atomic nucleus. Nonetheless, it offers an extraordinarily precise and concise description of certain phenomena. The magnetic properties of the electron, for example, are correctly described to 10 decimal places—the present limit of experimental accuracy. QED is generally regarded as a paradigm for a more complete theory which can include a description of the strong and weak nuclear forces.

Let me turn to the strong nuclear interactions. The atomic nucleus is known to be made up of more fundamental particles called nucleons. There are two species of nucleons: protons, which are the nuclei of the simplest atoms, those of hydrogen; and neutrons, which are unstable particles that can, however, survive when they are within a nucleus. The proton carries an electrical charge which is precisely equal and opposite to the charge of the electron. Opposite charges attract. So it is that a proton combines with an electron to form an atom of hydrogen. The neutron, like the hydrogen atom, is electrically neutral. Neutrons and protons are very small particles compared to the atoms they help to form. Once it was thought that they were truly pointlike and fundamental structures. Protons, neutrons, and electrons were generally believed to be the ultimate building blocks of matter.

All this has changed. The nucleon is now known to be a composite system made up of three quarks. It is the quarks that now seem to be the ultimate constituents of nuclear matter. Two kinds of quarks, called "up" quarks and "down" quarks, are needed to construct nucleons. The electric charges of the quarks are fractions of that of the electron, normally taken as "unit charge": "up" quarks carry ⅔ of a unit of charge and "down" quarks −⅓. A proton is a composite system containing two "up" quarks and one "down" quark; a neutron contains two "down" and an "up." Particles containing three "up" quarks or three "down" quarks are very short-lived, but they have been detected in the laboratory.

The fundamental interaction among quarks which binds them three at a time to form nucleons is known as quantum chromodynamics, or QCD. Although QCD and QED describe very different kinds of forces, there are certain profound similarities between them. Both are quantum field theories of the kind known as "gauge theories." QED leads to interactions between charged particles such as electrons and nuclei; it is QED that binds them together to form atoms. In a similar way, QCD leads to interactions which hold quarks together to form nucleons. The quantity analogous to the electrical charge of QED is called "color" in QCD. Quarks have color. More specifically, each "flavor" of quark (up or down) comes in three colors (red, green, blue, say). The nucleons, however, are "colorless." They contain one quark of each color. When these loose remarks are translated into a precise mathematical language, we discover why it is that *three* quarks—not two or four—can bind together to form an observable particle. We also understand why it is that a free quark cannot be produced, that is, one that exists alone. A quark is colored, and only colorless systems can exist as isolated particles. Quarks exist, but only as constituents of the particles they make up.

In the past few years, there have been a number of successful experimental tests of QCD. It now appears that we have a correct theory of the strong interactions, just as QED is a correct theory of electromagnetic interactions; though with QCD we cannot yet make predictions about nuclear processes that are accurate to ten decimal places.

I now turn to the weak nuclear interactions—those responsible for the process of nuclear beta decay. I have mentioned that a neutron on its own is unstable. After a few minutes, it decays into three stable particles: a proton, an electron and an antineutrino. (Neutrinos and antineutrinos are massless or nearly massless particles with neither strong interactions nor electromagnetic interactions. Ghostlike, they can travel through kilometers of matter without suffering a collision.) Until very recently, there existed no theoretically satisfactory theory of weak interactions; but now there does.

Each type of force is carried by a special kind of particle. Electromagnetism is mediated by photons: the force between two charged particles can be thought to be caused by the exchange of these photons. Similarly, the strong interactions are mediated by "gluons." Unlike photons, gluons cannot be seen directly because they, like the quarks, have color. However, there is good, but indirect, experimental evidence, for their existence. Since the 1930s, physicists have suspected that there is a special particle to mediate weak interactions: the "intermediate vector boson," or W particle.

The process of beta decay was supposed to proceed stepwise: first a neutron emits a W^- and becomes a proton; then the W^- becomes an antineutrino and an electron. Although this gives a useful and predictive picture of weak interactions, it is not a self-consistent picture.

In the past decade, it was realised that a sensible theory of weak interactions required the unification of weak and electromagnetic interactions. An "electroweak theory" has been developed that has had many triumphant experimental successes. Most important, it predicted the existence of a new class of physical phenomena called "neutral currents." These have been seen, and they agree in quantitative detail with what the theory predicts.

The electroweak theory, like QCD, is a gauge theory. So all three of the fundamental elementary-particle forces can now be included in a self-consistent and highly predictive quantum field theory. This "standard theory" offers an apparently correct and complete description of *all* phenomena, excepting gravitation. Yet, from an aesthetic point of view something is missing.

QED is a theory which involves only one arbitrary parameter: the strength of the electric repulsion between two electrons. The "standard theory" involves *seventeen* such arbitrary parameters. This is clearly too large a number to be truly fundamental. Moreover, the standard theory does not explain the fact of "charge quantization": that the electric charge of the electron is *exactly* equal and opposite to the charge of the proton.

Grand unified theories attempt to overcome these setbacks. More particularly, the theories take into account the observation that the fundamental quarks and leptons (particles *not* made of quarks) seem to come in simple "families." The first "family" consists of: the "up" quark, with charge ⅔; the "down" quark, with charge $-1/3$; the electron, with charge -1; and the neutrino, with zero charge. The first family is all that is necessary to operate "Spaceship Earth" and to fuel the Sun. However, there seem to be *three* such families of fundamental particles. The members of the other two families are the constituents of unstable particles that are produced by cosmic rays or at particle accelerators. Each family consists of two quarks and two leptons, with the same electric charge structure. We do not understand why in nature there are three families and not just one.

In grand unified theories, each of the families is regarded as a single structural unit. Particles as different as quarks and leptons are grouped together. The electrical charges of quarks and leptons hint that this is sensible: if we remember that quarks come in three colors, we see that the sum of the charges of the particles in each family is exactly zero. I must say a word on what it means to "put together" dissimilar particles or dissimilar forces.

The degree of symmetry of a system often depends upon temperature: usually there is more symmetry at high temperature than there is at low temperature. A liquid, for example, is "isotropic": all directions are equivalent. When it cools, it may form crystals. These are anything but isotropic, but possess a well-defined symmetry which is less than total isotropy. Permanent magnets offer another example. The dissymmetry reflected by the existence of separate north and south poles disappears when the magnet is heated and loses its magnetism.

Thus it is that particle physicists regard the observed dissymmetries of nature as an artifact due to the prevailing low temperature of the universe. At higher temperatures, the innate symmetry of the universe is revealed more clearly. Unfortunately, "high" here means very high indeed. There are three domains:

HOT. Here I mean temperatures in excess of 10^{29} degrees Kelvin, or energies of 10^{16} GeV. Under these conditions, there is complete symmetry. Quarks and leptons are massless and not distinguishable from one another. Strong, weak, and electromagnetic interactions are evidently equivalent parts of a grand unified theory. These temperatures are today quite inaccessible. They were achieved only in the earliest moments of the Big Bang. Since then, the universe has congealed, thereby losing its symmetry.

LUKEWARM. This is the region between 10^{29} degrees Kelvin and 10^{15} degrees Kelvin, or between 100 GeV and 10^{16} GeV. The lower range of energy will soon be accessible to accelerators which are now being built in Europe and the U.S. At these temperatures the symmetry between weak and electromagnetic interactions should be evident. This is the range of energy in which the unified electroweak theory can be put to precise test.

COOL. At lower energies, all the symmetry of the universe is lost except "color," which is at the root of strong interactions, and "charge," which generates electromagnetism. Weak interactions are observed to be weak because everyday energies are so much smaller than the electroweak scale of 100 GeV.

The simplest grand unified theory is based on the mathematical group known as SU(5). This is the unique unified theory in which the basic families of particles are simply the known quarks and leptons. There are exactly 24 particles in this theory whose exchange produces forces: the *photon*, which mediates electromagnetism, and is both massless and observable; the *three weak intermediaries,* which cause the weak interactions, and are massive particles (W^+, W^- and Z°) which weigh about 100 GeV (in energy terms) and are observable—they should be seen with the aid of the next generation of accelerators; the *eight gluons,* which mediate the strong "chromodynamic" interactions, and, being colored, cannot be observed in isolation, although their indirect "footprints" are seen quite clearly; *twelve more particles,* which are very massive (about 10^{16} GeV), very shortlived, and quite unobservable. These last particles lead to a new force 10^{28} times weaker than ordinary weak interactions. This new force can convert quarks into leptons, and so can lead to the decay of protons—the most dramatic prediction of grand unified theories. All matter therefore decays, although with a very long half-life. It should take the average proton 10^{31} years to decay, give or take a factor of 10.

In a ton of matter, the theory predicts that one proton should decay in each decade. Although this seems to be a very small effect, it can be detected. Experiments are being deployed in underground mines involving hundreds or thousands of tons of material: in India, Utah, Ohio, Minnesota, and perhaps under the Alps. Within a year or two we should know whether or not the proton does decay, and whether or not it is true that "diamonds are forever."

Not only does grand unification predict the death of all matter, but it may provide an explanation for the birth of matter. Scientists have long wondered why the universe contains matter but no antimatter, but only today are we finding the answer. In the very young universe, the force which produces and destroys protons was much stronger than what it is today. It is believed that the same force which will lead to proton decay was once responsible for the production of matter in the first place.

Grand unification has only a few experimental tests. It predicts the value of a parameter which I invented, but which is known as the Weinberg angle. Five years ago, this prediction was in serious disagreement with experiment. With more precise experiments alone, the experimenters have changed their tune. The experimental value of the Weinberg angle is today in excellent agreement with the value that is predicted by SU(5). We shall soon see whether or not the proton decays, as SU(5) says it must.

Sadly, the result of the proton decay experiments is that protons live much longer than SU(5) says they do. The simplest grand unified theory has been proven to be wrong. Yet, these experiments led to a serendipitous discovery. The proton decay detectors were installed just in time to detect a burst of neutrinos coming from the supernova of 1987. Instead of confirming my theory of proton decay, these experiments proved that astrophysicists really do understand what a supernova is.

The virtues of grand unification are almost without number. It is obviously a precursor to the fulfilment of Einstein's dream of unifying *all* the interactions. It explains the fact that the photon is a massless particle. In the standard (not unified) theory, the photon could have had mass, but chooses not to. In a grand unified theory there is no such possibility. The fact of charge quantization is likewise explained. The equality of proton charge and positron charge is forced upon us, as well as the curious fractional charges of the quarks and the zero charge of the neutrino. In SU(5), there simply is no other way. Each family of quarks and leptons is a well-defined representation of SU(5) with properties that *must be* just as they are *seen to be*. We even understand why the strong interactions are so much stronger than electrodynamics. When the universe was young, the interactions were of the same strength. In the cold, ''crystalline'' universe of today, they have come apart in a calculable and well-understood manner.

Perhaps the greatest virtue of grand unification is its prediction of a new class of observable phenomena. In SU(5), or in its simpler elaborations, the proton (and hence all nuclear matter) must decay with an observable lifetime. Indeed, the predicted lifetime is only slightly longer than the established lower limit of 10^{29} years. (Incidentally, we feel it ''in our bones'' that the proton lifetime is longer than 10^{16} years. Were it so short, we would be killed by the radioactivity produced by decaying matter within our own bodies!) This prediction comes at a very fortuitous time. We may be approaching the end of the line in our pursuit of higher and higher energies. The Large Electron-Positron (LEP) machine will cost Europe over $500 million and use a great deal of electric power. Larger machines may be built, but clearly not very much larger. Surely we can never achieve an energy wherein grand unification is apparent: such a machine would consume more power than the Sun produces. How fortunate it is that nature has provided us at least two windows into the world of high temperature: the universe as a whole, which was baked at such temperatures during birth, and the possible decay of matter. Why is there matter? And, does it decay?

Grand unification suggests (but does not quite demand) the existence of other forms of matter that may be studied away from the high-energy frontier. One such form is magnetic monopoles, particles first invented by Paul Dirac decades ago. They are to magnetism as electrons are to electricity—isolated magnetic north or south poles, unlike magnets or other known magnetic systems which bear north and south poles together. In grand unified theories such particles should exist, not as fundamental particles, but as ''topological condensates'' or undoable knots in the fabric of quantum fields.

The monopoles of grand unified theories are heavy, weighing about 10^{16} GeV. Their mass, converted into energy of motion, is that of a bus at high speed! Monopoles have been searched for, but never monopoles like these. So heavy are they that when placed on a table, they would fall through to the core of the Earth. Possibly, monopoles are incident on Earth in cosmic rays, perhaps as many as one monopole per square meter per day.

Grand unification suggests that neutrinos should have measurable masses and be subject to oscillations. This means that they may change their identities as they move along. A number of experiments can be interpreted in terms of such effects. There is a well-known deficit in the number of solar neutrinos incident upon us. Perhaps these neutrinos have changed their spots, becoming undetectable to us. Recent accelerator experiments at CERN and a reactor experiment in the U.S. suggest neutrino mixing. And, a recent Soviet experiment claims to observe a non-vanishing mass of the neutrino. All of this is fraught with significance, not only for physics but for cosmology. Some astrophysicists are beginning to believe, not only that neutrinos have mass, but that neutrinos are the dominant form of mass in the universe. Protons and neutrons are the rare and exotic materials of which we and our stars are fashioned.

Grand unification may, as I have said, turn out to be merely fools' gold. But, it will soon be tested. It offers the promise of exciting discoveries in particle physics, well away from the expensive high-energy frontier. It deals with the birth of matter, the death of matter, and the fundamental nature of the universe today. What more can we ask of a mere theory of physics?

On the Way to a Unified Field Theory

Most physicists have an unshakable faith in the underlying simplicity of nature; it is one of our most powerful guiding principles. Time and time again, this blind faith has been vindicated. Many of nature's bewildering tricks have been explained. What could be more different than magnetism, electricity, and light? Yet, in the nineteenth century, James Clark Maxwell showed that these phenomena were simply different manifestations of the same fundamental laws. He described all these, as well as radio waves, radar, and radiant heat, by a unique and elegant system of equations. Maxwell's electromagnetism is the operative force controlling everything that we see, hear, smell, feel, or taste.

Thus began the quest for a unified theory of all of the forces to which matter is subject. Albert Einstein knew of only two fundamental forces. Gravity keeps the planets in orbit about the sun and our feet upon the ground. Electromagnetism keeps atomic electrons in orbit about the nucleus, allows atoms to combine together into molecules, and ultimately explains why we do not fall through the ground. Einstein attempted, without success, to build a unified theory of these two forces. One reason for his failure is that there are more forces in heaven and earth than were dreamt of in Einstein's philosophy.

We have learnt that two additional forces are required to explain the behavior of the atomic nucleus. These small and heavy kernels within atoms are made up of still smaller constituents called neutrons and protons. It is the strong nuclear force that holds these particles together to form the nucleus. In a certain kind of radioactivity, beta decay, a neutron within a nucleus emits two particles (an electron and an antineutrino) and becomes a proton. It was Enrico Fermi who interpreted this process in terms of a fourth force called the weak nuclear force.

A truly unified theory would explain all four of these forces. This ambitious challenge has not yet been met. What is being developed is a unified theory of the strong, weak, and electromagnetic forces: everything except gravity. These three fundamental forces of the microworld are described by what physicists call gauge theory, in which the basic particles are of two kinds, which we call force particles and matter particles. Complex physical phenomena result from the iteration of a primal act of becoming, the emission or absorption of a force particle by a matter particle. For example, one matter particle may emit a force particle which is then absorbed by a second matter particle. This exchange produces a force between the two matter particles.

Photons are the force particles which mediate electromagnetism. The force between an electron and its nucleus results from the exchange of photons. Photons are observable particles—we see them all the time as light.

The force particles which are responsible for weak interactions are known as $W^{\pm}$ and Z°. Unlike the photon, these particles are very heavy—almost 100 times heavier than the proton, according to theory. No existing particle accelerator is big enough to produce these particles. We deduce their existence indirectly. When a neutrino collides with a neutron, occasionally a W is exchanged. The neutrino becomes an electron, and the neutron becomes a proton. A variant of this mechanism leads to the much-studied process of nuclear beta decay. The Z° particle mediates a class of weak phenomena which was first observed by a European collaboration at CERN in 1973. A partially unified theory of weak and electromagnetic interactions, which was formulated in the 1960s, successfully predicted the existence of neutral currents and earned the 1979 Nobel Prize for Abdus Salam, Steven Weinberg and this author. The ultimate test of the electroweak theory will be the production and observation of the Z° particle in the laboratory. Accelerators which can accomplish this feat are in construction in Europe, the United States, and the Soviet Union. The CERN collider, in Geneva, Switzerland, was the first of these great new machines to be built.

Strong interactions are thought to be mediated by a third class of particles called *gluons*. These are much like photons, with two big differences: There are eight distinct kinds of gluons, but only one kind of photon. Moreover, the strong force (as its name suggests) is much stronger than electromagnetism. A consequence of this is that gluons are permanently confined within nuclear particles. You cannot make a gluon flashlight the way you can make a photon flashlight.

All told, there are exactly 12 different kinds of force particles: eight gluons, one photon, $W^{\pm}$ and Z°. We must also know what are the different kinds of matter particles. Some kinds of matter particles are involved with the strong interactions. They can emit and absorb gluons. These matter particles are called quarks. Matter particles which have no strong interactions are called leptons.

Today, physicists believe that there exist in nature six different species of quarks. Two of them, called up quarks and down quarks are the fundamental constituents of atomic nuclei. The protons and neutrons, which are the constituents of the nucleus, are not really fundamental particles at all. They are made of quarks. The proton is made of two up quarks and one down quark held together by gluons, while the neutron contains two down quarks and an up quark. It follows that quarks carry electric charge Q which is a fractional part of the proton's charge. For the up quark $Q = 2/3$, while for down quarks $Q = -1/3$.

The remaining species of quarks are not essential constituents of nuclear matter. They are much heavier than up and down quarks, and they decay into lighter quarks by virtue of the weak interactions. Strange particles, which is to say, particles containing strange quarks, have been studied since the 1950s. They have lifetimes of about 10^{-10} seconds. Charmed particles were discovered in the 1970s. They are particles containing charmed quarks. Much heavier than strange particles, charmed particles also have much shorter lifetimes. Ingenious experiments performed at CERN, and involving a largely Italian collaboration of experimenters, succeeded in 1979 to show that charmed particles live about 10^{-13} seconds.

A new particle called upsilon was discovered in an experiment done at Fermilab in 1977 by an American collaboration. It is now known to contain a fifth kind of quark, the so-called bottom quark. Well over a dozen particles containing one or two bottom quarks have already been discovered.

Of the five known quark species, two (up and charm) have electric charge $Q = 2/3$, while three (down, strange, bottom) have electric charge $Q = -1/3$. Theoretical arguments suggest that there should exist equal numbers of $Q = 2/3$ and $Q = -1/3$ quarks. The existence of a sixth quark with $Q = 2/3$ is expected, and the sixth quark is called the top quark. In an international race, European and American experimenters are actively searching for evidence for the top quark, which may be the last of the quarks.

Leptons are the matter particles which do not possess strong interactions. The most familiar lepton is the electron, which was discovered in 1895. It has electric charge $Q = -1$, opposite but exactly equal to the charge of the proton. Two heavy $Q = -1$ cousins of the electron are known. The muon weighs about 206 times more than the electron and was first observed in 1938. The tau lepton weighs about 17 times more than the muon, and was discovered at Stanford, California, in 1976.

Each of these charged leptons is associated with its own uncharged $Q = 0$ lepton called a neutrino. Neutrinos were generally thought to be massless particles, but recent experiments and astrophysical observations suggest that they do have masses. Indeed, there are so many neutrinos in the universe that, collectively, they may account for the bulk of its mass, even though an individual neutrino is very light.

Why are there three families, when a respectable version of the universe (it would appear) could be fashioned by using only one? Is there a fourth family waiting to be discovered at higher energy? These are questions which we cannot yet answer. Why do the quarks and leptons group themselves into families? This question is answered in the framework of what is called the grand unified theory. The two different types of matter particles are inextricably entwined with one another, and the periodic table of quarks and leptons is forced upon us.

Many more of the most puzzling questions about particle physics are answered in the framework of grand unification:

1. Strong, weak, and electrodynamic forces are seen to be avatars of the same fundamental mathematical system.

2. Why is the photon exactly massless? In the grand unified theory, this simply must be.

3. Why are the strong interactions so much stronger than electromagnetism?

4. Why are neutrinos electrically neutral and nearly massless?

5. Why is the electric charge of the proton exactly equal in magnitude to that of the electron?

6. Why is matter stable? The answer is that it is not!

7. How did the matter in our universe originate?

A few years ago, we could not even dare to ask such difficult questions. Clearly, physicists have come a long way in a short time. Eddington once prophesied that a time would come when physics (like chemistry before it) would become complete and hence uninteresting. The time is not yet, but soon it may be.

Tangled in Superstring

WITH BEN BOVA

From the grandest spectacles in the heavens to the most minute twinges of the ultimate particles, science, in its varied disciplines, is called upon to explain all the phenomena of nature. Cosmologists deal with the biggest questions: the birth of the universe and the origin and development of its billions of galaxies. Astronomers work their way down the cosmic ladder to things as small as our solar system. Geologists are concerned with all the nooks and crannies of our planet, while biologists study the things on Earth that creep and crawl and swim and fly and infect one another, from wee viruses to the great whales they attack. Next come chemists and most of the physicists, whose job it is to explain the bulk properties of matter in terms of the minute atoms of which everything on Earth is made. They will tell you why copper is red, why the sky is blue, how a candle burns, and what makes dew.

The still smaller world inside the atom attracts the attention of two distinct groups of physicists, working in increasingly tiny domains. First there are the nuclear physicists, who study the atom's central core—the key to nuclear weapons and nuclear power, and the marvelous nuclear furnace we call the sun. Then there are the elementary-particle physicists, who study the constituents of atomic nuclei—neutrons and protons and all the other particles, once thought to be elementary, but now known to be made up of quarks—as well as the forces that govern interactions between the various particles: electromagnetism and the strong and weak nuclear forces. With the most powerful accelerators in use today, in which particles collide with a force of two trillion electron volts, it is possible to generate things as small as the W and Z particles, a hundred *million* times smaller than the atom.

Theoretical physicists speculate on the existence of even smaller entities, things so small that no conceivable accelerator could spot them. In various grand unified theories, the three particle forces are seen as variations on one simple force, a force whose workings can be discerned only on a very small scale, because it depends on the existence of very heavy and incredibly tiny new particles, a *trillion* times smaller than the W and Z particles.

At distances 100,000 thousand times smaller still, 10^{-33} centimeter, we finally reach a natural end point. Quantum gravity (the gravity that physicists believe must exist in the subatomic realm but that cannot be detected) becomes the dominant force, and there is no agreed-upon theory that can deal with it. This Planck length defines, for now, the very bottom of the cosmic ladder. Supposing that each rung corresponds to an increase in size by a factor of 10, there are 52 rungs from the Planck length, up through the world of inches and feet, to the size of the entire visible universe.

Along most of the ladder, scientists know pretty well what they are doing and where they are going. Chemists, for example, agree that their discipline is closed, in the sense that no fundamental principles remain unknown. This is not to say that there are no hard problems to be solved, nor discoveries to be made, but that the basic rules are fixed: atoms interact with one another according to the laws of quantum mechanics, laws whose validity has been established beyond a reasonable doubt.

It's at the extreme ends of the cosmic ladder that the wild things are. Cosmology and particle physics are open-ended disciplines; we really don't know all the rules of those games. How and why did the universe evolve to become as it is, and how will it behave in the distant future? Will it continue to expand and cool indefinitely (the big chill) or will it one day reverse itself, shrink, and eventually implode (the big squeeze)?

The story is the same at the other end of the ladder. The so-called standard theory of particle physics describes the 17 known particles and their behaviors. These particles are of two kinds: fermions, or particles of matter (including protons, neutrons, electrons, and the various quarks), and bosons (photons, gluons, and the W and Z particles), which carry the electromagnetic, strong, and weak forces that govern interactions between fermions. But are these truly the ultimate forms of matter? No experimental data contradict the standard theory, nor do they indicate the existence of any structure lying outside its domain. Yet the standard theory leaves many of the juiciest questions unanswered—the most important of which concerns how quantum gravity works at the bottom of the cosmic ladder. So elementary-particle physics has reached either a curious crossroads or an insurmountable impasse.

Gravity is the dominant force at the higher rungs of the cosmic ladder, where its weakness is made up for by the sheer number of atoms exerting force in concert. It binds the stars into galaxies, and it keeps the sun intact and its family of planets firmly in orbit. It glues the moon to earth and holds the oceans and atmosphere in their proper places.

But when it comes to things a few kilometers in size, the force of gravity must yield to electromagnetism. Gravity tries, but it cannot stop a tree from growing or a mountain from forming. Each time we arise in the morning, each time a spring flower blooms, the supremacy of electromagnetism is displayed. Its majesty ranges from things the size of continents to elephants and microbes and to their component atoms.

Farther down the cosmic ladder, electromagnetism is superseded by the strong force. In its limited domain, this force is all-powerful; it holds protons and neutrons together to form the atomic nucleus, much against the inclination of electromagnetism, which would just as soon blow the nucleus apart. The quarks that compose protons and neutrons are also bound by the strong force. But another four rungs down the ladder, the strong force gives way to the weak force, which causes particles and nuclei to disintegrate spontaneously, emitting energy.

We have reached the rung on the ladder corresponding to the highest-energy accelerators at our disposal, with another 17 rungs leading down to the Planck length. Our knowledge of the physics of this domain is scanty and speculative. Because of the spectacular successes of the standard theory in explaining all the subatomic interactions we can detect, and because of the lack of surprising new discoveries in recent years, some theorists have suggested that nothing at all of great interest takes place there. According to this philosophy of despair, there are no surprises awaiting us—no new particles, no new forces, and no interesting phenomena to observe. Have we really entered upon a great desert that extends from the energies of two trillion electron volts, all the way to the Planck length, and perhaps beyond?

I doubt that we have. A few grains of sand do not necessarily indicate a desert, but a pessimistic philosophy can make the desert a self-fulfilling prophecy. Fortunately for science, several accelerators will soon be completed, and we will be able to see for ourselves whether we face a Sahara or the shores of an uncharted ocean.

In the end, we must confront the physics at the bottom of the ladder, where quantum gravity becomes important, even though we cannot build accelerators powerful enough to explore this domain. The universe is—or, more precisely, was, at the time of its birth, 15 billion years ago—the most powerful accelerator ever. Perhaps, if scientists are ingenious enough, the unseen relics of the Big Bang, which are thought to be scattered about us, will provide the key to the physics of ultrahigh energies.

But many of the best and brightest young physicists have chosen a different, entirely theoretical, approach to understanding matter—an approach not based on experimentation and whose constructs never have been demonstrated. *Superstrings* is the name of their game, a synthesis of some of the most bizarre notions ever put forward. According to this new religion, space has nine dimensions, not just the three we see. Six of them are curled up into a tiny ball whose radius is the Planck length and, so, are far too small ever to be noticed by our big and clumsy species. Superstring theory identifies particles, not as pointlike structures, but as tiny loops of string. There is only one kind of string, and the wiggles of the loop in nine-dimensional space determine whether it acts as a quark, or a photon, or some other particle.

The popular literature brims with glowing articles about superstrings and the consequent theory of everything. Its proponents are acclaimed as the new successors to Einstein, and string physicists everywhere are convinced that they are on the verge of the ultimate breakthrough, a complete understanding of the nature of the physical universe. What's all the hoopla about?

Superstrings, for the first time ever, appear to present us with a theory of quantum gravity, and I've got to give them a brownie point for that. What's more, superstring theory seems to be unique, or almost unique, in that it does away with the need for measurement. In principle, the electron mass, the proton mass, the strength of the electromagnetic coupling, and all other measurable characteristics of matter are calculable from the theory without resort to experiment. The rules of particles and forces that make up the standard theory, our successful description of the low-energy world, are expected to emerge logically from superstring theory as necessary consequences.

But superstring physicists have not yet shown that their theory really works. They cannot demonstrate that the standard theory is a logical outcome of string theory. They cannot even be sure that their formalism includes a description of such things as protons and electrons. And they have not yet made even one teeny-tiny experimental prediction. Worst of all, superstring theory does not follow as a logical consequence of some appealing set of hypotheses about nature. Why, you may ask, do the string theorists insist that space is nine dimensional? Simply because string theory doesn't make sense in any other kind of space.

Is there gold at the end of the road or merely a morass of ever more abstruse mathematics? Even the most ambitious string advocates believe that it will take decades before we have learned enough to tell for sure to make any experimental predictions at all. Meanwhile, the historical connection between experimental physics and theory has been lost. Until the string people can interpret perceived properties of the real world, they simply are not doing physics. Should they be paid by universities and be permitted to pervert impressionable students? Will young Ph.D.s, whose expertise is limited to superstring theory, be employable if, and when, the string snaps? Are string thoughts more appropriate to departments of mathematics, or even to schools of divinity, than to physics departments? How many angels can dance on the head of a pin? How many dimensions are there in a compactified manifold, 30 powers of ten smaller than a pinhead?

The last great experimental surprise in particle physics, the discovery of a third family of particles, took place during the celebration of the U.S. bicentennial. The tau lepton was found in California, and the bottom, or "beauty," quark showed up in Illinois. More than a decade has passed. A conference entitled, "The Fourth Family of Quarks and Leptons," was held in Santa Monica, though there is not a bit of experimental evidence, or theoretical motivation, to suggest that more than three families of such particles exist. Particle physicists are so desperate to have something to talk about that they are playing "Let's Pretend."

History is on our side. Every few years, there has been a dramatic discovery in fundamental physics or cosmology—well more than a hundred of them since Newton's day. Can anyone really believe that nature's bag of tricks has run out? Of course not.

Grand Unification and the Future of Physics

The grand unified theory was invented in 1974. In it, all of the forces of the microworld are described by a single unified system of equations. Quarks and leptons are seen to be equal partners in the scheme of things. Even the birth and death of matter is explained.

At first, no one believed the theory because it made a prediction which was in disagreement with experimental data. An observable parameter (the weak mixing angle) was predicted to be 27 degrees, and observed to be 38 degrees. More careful experiments revealed that the original experiments were wrong. Today, the experimental result is in excellent agreement with the prediction of grand unified theory. The moral is clear: When a theory is so elegant and so beautiful that is *has* to be true, then it *will* be true.

The force particles associated with electromagnetic, weak, and strong interactions are very different, as shown in Table 1.

TABLE 1. Force Particles

Type of Force	*Number of Intermediaries*	*Mass of Force Particle*	*Can They Be Isolated?*
Electromagnetic	1	Massless	yes
Weak	3	Heavy	yes
Strong	8	Massless	no

How can such different entities be put into a single unified theory? We can approach this difficult question by means of analogies.

Consider the paradox presented by a magnet, which has both a north pole and a south pole. The two ends of the magnet are very different, yet the laws of physics are perfectly symmetrical, and do not distinguish one side of the magnet from the other. However, if the magnet is heated in an oven, it will lose its magnetism. The underlying symmetry of nature is restored, and the two ends of the magnet become identical. As another example, consider water vapor. It is isotropic, homogeneous, invisible, and quite without beauty. No direction is distinguished—it displays the symmetry of our laws of physics. Let it cool, and the symmetry destroys itself. Lesser symmetry with more structure is revealed: a snowflake appears. Magnets and snowflakes are examples of a common physical phenomenon: symmetry breaking.

Thus it is that the symmetry of our grand unified theory is manifest only at very high temperatures—temperatures so high they cannot be obtained in the laboratory, or even in an exploding star. Only once in the history of our universe were the symmetries of grand unified theory explicitly realized: at the moment of its birth, in the primordial Big Bang. The wondrous variety of natural phenomena—strong, weak, and electrodynamic interactions, quarks and leptons, appear like snowflakes in the cold and crystalline universe which we now inhabit.

Our theory of strong, weak, and electromagnetic interactions is based upon a recondite branch of abstract mathematics, the theory of Lie groups. The classification of these structures was worked out in the nineteenth century by the Norwegian mathematician Sophus Lie and the French mathematician Eli Cartan. They showed that a wide class of Lie groups could be expressed in terms of more fundamental structures called Simple Lie groups. The search for a grand unified theory is the search for a simple Lie group which contains a description of the known force particles (photons, $W^{\pm}$ and Z°, and the gluons), and of the known matter particles (quarks and leptons). If our periodic table of matter particles is correct, in the sense that there are no missing columns, then the choice of a simple Lie group is unique. Our grand unified theory must be based on the group known as A_4 or SU(5).

The force particles of strong, weak and electromagnetic interactions are exactly 12 in number. Yet, the group SU(5) involves 24 such particles. Grand unified theories involve the necessary existence of a new kind of force, mediated by the 12 additional force particles. The reactions which are mediated by this new force have not yet been detected. They are responsible for the eventual decay of all matter. The new interactions are very weak, so that the lifetime of matter is very long. Diamonds may not be forever, but almost. In a ton of matter, only one nuclear particle should decay in a decade. Nonetheless, it is possible to search for this process. In a deep mine, which is shielded from cosmic rays, one can look for the occasional decay of a proton or neutron. Such experiments are being launched under the mountains between France and Italy, in Indian gold mines, and in mines in the United States. The central prediction of grand unified theory—that all matter is radioactive—will soon be put to a decisive test. Soon we shall know whether or not "diamonds are forever."

Just as grand unification predicts the ultimate disappearance of all tangible matter, it also helps to explain how matter was formed in the very young and hot universe. It was Andrei Sakharov, perhaps the greatest living Soviet physicist, who sketched the scenario for the birth of matter in 1967. Three things were necessary in order that matter could be created, he argued. There had to be a Big Bang—a tremendous explosion marking the beginning of time. This theory is now an accepted feature of our standard scientific cosmology. Secondly, the forces of nature had to define the arrow of time. This was established by the experimental work of Fitch and Cronin, who received the Nobel Prize for their epochal work in 1980. Finally, it was necessary that a mechanism existed for the decay of matter—the ingredient supplied by the grand unified theory. Recent theoretical work has shown that Sakharov's vision of the birth of matter makes sense in the framework of the new synthesis.

The grand unified theory resolves a number of old puzzles in particle physics. It explains why there exist both quarks and leptons in nature, and why they occur together in families. The structure of the periodic table of quarks and leptons is forced upon us. Why is the electric charge of the proton observed to be exactly minus that of the electron? This is neither a mystery nor a coincidence in the grand unified theory—it is the way things must be. Why is the photon exactly massless? In a theory which is not unified, this does not have to be true. In a unified theory, it does. Why are strong interactions so much stronger than electromagnetism? In a unified theory, both interactions are of the same strength at very high temperatures. In the cold universe that we inhabit, they have drawn apart. Again, our theory is a success in that things *must be* as they are observed to be.

Attractive as grand unification is, it must be put to a precise experimental test. Three crucial predictions must be verified. The first test involves the structure of the weak neutral currents. Grand unification demands that a certain observable parameter (called the "weak mixing angle") takes a definite numerical value. Data that is now available is in agreement with the prediction to within 10 percent. Better data is needed to test the theory. The second test is the prediction of proton decay. Not only must the proton decay with a predicted lifetime, but it must decay in a predicted manner. Many experiments will have to be done to verify these predictions.

The third test of grand unification is somewhat different. Particle physicists have become accustomed to the discovery of new and exciting phenomena as they reach to higher and higher energies. Every new accelerator, almost without exception, has produced its share of surprises. Grand unification predicts that the end of the high-energy frontier is approaching. The W and Z particles must be observed. The top quark remains to be found. Particles containing top quarks and bottom quarks must be studied. Just possibly, there may remain a fourth family of quarks and leptons at higher energies. Beyond these details, grand unification demands that there are no more surprises. It predicts the existence of a "great desert" at high energies, where there will be little of interest to study.

Some physicists are appalled by the prospect of a vanishing high-energy frontier. Surely there will be new wonders awaiting the next generation of accelerators, they argue. Perhaps they are right, but if we do not build the larger accelerators (like LEP, the Great European Dream Machine), we shall never know. On the other hand, questions of cost, power, and limited resources suggest that we cannot go on forever building larger and larger atom smashers. Perhaps in the foreseeable future, we will have learned all we need to know of the world of high energies. We will have all the clues we need to fashion the one and true theory of physics. The incipient end of the high-energy frontier should be regarded as a triumph, not as a tragedy.

Accelerators have been our principal tool for only a few decades. In fact, many of the great discoveries in particle physics did not depend upon such artificial aids. And, I am sure that many great discoveries remain to be made which are far away from the high-energy frontier. We have mentioned proton decay experiments, which need no accelerator at all, but use massive underground detectors. Many great surprises may lie just below our feet. The study of neutrinos is another discipline which does not need ever larger accelerators. These ghostlike particles can travel freely through the earth. They are produced by the sun, by the stars, by radioactive materials, by nuclear reactors, by cosmic rays, and by small and large accelerators. Do neutrinos have mass? Do they change their identities as they move along? Only recently have we begun to address these important questions. The suspicion is growing that neutrinos may comprise the dominant form of matter in the universe. *They* will determine whether or not the universe is closed. Ordinary matter, such as we and our stars are built of, may be but a minor contaminent of a universe which is mostly neutrinos.

Freed of our high-energy obsession, we may return to a study of the normal stuff of the universe. The last variety of tangible matter discovered on earth was the element rhenium, in 1923. Perhaps there are other strange things cohabiting with us. Isolated quarks with fractional charges have been avidly searched for. Our theory says that such searches will necessarily fail. But, theories can be wrong.

Perhaps there is a new level of interesting physics at inaccessible energies. This may lead to the existence of new stable particles on earth. One example is called the ''magnetic monopole.'' All magnets that have been seen have both a north pole and a south pole. Could there exist a particle which has only a north pole, or is this a senseless concept, like a piece of string with only one end? The English physicist Paul Dirac invented the magnetic monopole 50 years ago. Nobody has found one yet, but many physicists are still looking.

Perhaps there exist superheavy atoms, atoms which weigh thousands of times more than they should. Particles like this have never seriously been looked for. However, nothing in our theory of physics says that such particles cannot exist. If they do exist, they may be of immense technological importance. They may be the key to the construction of controlled nuclear fusion—the solution to the energy crisis. They may be partly responsible for the mechanism that powers the sun, about which we may think we know far more than we do. Here is another direction for experimental research that lies far from the high-energy frontier.

Finally, there is gravity, and the quest for a truly grand synthesis. Galileo, Newton, and Einstein all believed in the fundamental unity of celestial and terrestrial phenomena. It is not enough to have a theory of the microworld; of strong, weak, and electromagnetic interactions. The grand synthesis must include a description of gravity as well. Gravity dominates the world of large phenomena—planets, stars, and galaxies. It also dominates the smallest phenomena, at distances of 10^{-33} cm. Our so-called ''grand unified theory'' is a bit like the Holy Roman Empire—neither Holy, nor Roman, nor an Empire. It cannot deal with things that are too large, or with things that are too small. It is a theory of the ''in-between.'' The greatest challenge of theoretical physics still remains with us.

THE PHYSICIST AND SOCIETY

SSC: *Machine for the Nineties*

WITH LEON M. LEDERMAN

Often in the history of elementary-particle physics, high energy has been the key to new discoveries. In the middle of the nineteenth century, experiments at energies of several electron volts led to the discovery of emission and absorption spectra, and to the realization that the atom is a structured system. In the early part of this century, X-ray experiments at energies in the keV range revealed the inner structure of the atom and led Henry Moseley to the concept of atomic number in 1913. The evolution of nuclear physics, from the discovery of the atomic nucleus itself in 1911 to the mature discipline of modern nuclear science, began with experiments at energies on the order of MeV, first with particles available from naturally radioactive sources, but later from small and specialized machines, such as Van de Graaff accelerators. Another factor of a thousand in energy was required to expose the mysteries of the subnuclear world. The Bevatron at The University of California at Berkeley, operating at a center-of-mass kinetic energy of 2.7 GeV, first produced and detected antiprotons in 1955. In 1977, the upsilon particle (sign of the fifth quark) was discovered at Fermilab, where the center-of-mass energy was 27.5 GeV—ten times greater than the energy available a generation earlier.

In these comparisons of available energy, the relevant criterion is center-of-mass energy, rather than beam energy. The largest beam energy now available worldwide is generated at the Fermilab accelerator, known as Tevatron II. Its beam energy is at present 800 GeV, but, because it operates in a fixed-target mode, the available center-of-mass energy is only 40 GeV. On the other hand, the CERN proton-antiproton collider in Europe has about 300 GeV in each beam; so it has 600 GeV available energy, which is precisely what allowed the discovery of the W and Z particles in 1983.

Not all great discoveries have depended on access to the largest possible energies. The J/ψ particle (see glossary), the tau lepton, the upsilon, and bare charm were all discovered in the U.S. during a time when the CERN Intersecting Storage Rings were clearly ahead in the high-energy sweepstakes. Skill, imagination, persistence and experience are vital research ingredients; a high usable luminosity also helps. It is for this reason that fixed-target physics is complementary to colliders. However, particle physics has reached the point where even good old American know-how cannot compete with the higher energies that colliders can make available. The CERN collider, having confirmed the central predictions of the electroweak theory, has certainly earned the Nobel Prize for European high-energy physics—the first in many years. But, more importantly, Europe has shown the way to the style of research required to address the open questions—the hadron collider furnished with integrated and sophisticated detectors.

The development of colliding-beam physics began at Stanford University, which operated an electron-electron collider in 1963. Electron-positron colliders followed at Frascati, Orsay and Novosibirsk. The CERN Intersecting Storage Rings, the first hadron-hadron collider, was commissioned in 1971. A series of electron-positron rings have operated at the Stanford Linear Accelerator Center (SLAC), Orsay, Frascati, Novosibirsk, Hamburg and Cornell University. However, with the CERN $p\bar{p}$ collider, the field will remain wide open to European research until Tevatron I becomes effectively operational at Fermilab. Table 1 summarizes the present and planned accelerator inventory.

We have come a long way in our search for an understanding obtained from what are now four generations of postwar accelerators. We have identified six quarks, which are the constituents of all hadrons, that is, all particles that interact via the strong force. We have evidence for six leptons. We have a partially unified electroweak theory and a very promising theory of the strong (color) force, quantum chromodynamics. The data support the idea that the forces are described by a quantum field theory obeying gauge invariance. We have now identified all the corresponding gauge bosons, the carriers of the forces: photon, $W^{\pm}$, Z^{0} and

TABLE 1. Glossary of High-Energy Physics Jargon

J/ψ	The compromise name for a particle discovered simultaneously at Brookhaven (J) and SLAC (ψ) in 1974, and which was soon interpreted as the bound state of a charmed quark and a charmed antiquark. The charmed quark was suggested by J. D. Bjorken and Glashow in 1964.
τ Lepton	The third charged lepton (after the electron and muon), discovered at SLAC in 1976.
Υ	The discovery of three closely spaced particle states at Fermilab in 1977 by Lederman and collaborators was soon interpreted as the bound states of a new quark, the bottom quark, b, and its antiquark. Together with the τ, this established the third generation of quarks and leptons.
Bare Charm	Since the J/ψ contains both charm and anticharm, the intrinsic properties of the quarks are hidden. A state containing a charmed quark with, say, an anti-up quark, exposes the quantum numbers of charm and represents bare charm. (The Υ is an analogous system containing bottom quarks.)
W, Z	The force carriers of the weak force, discovered at CERN in 1983.

the gluons. All of this constitutes a powerful synthesis called the standard model (see Table 2).

It has been said that the standard model offers a complete and correct description of all observed phenomena on earth, and perhaps in the universe. Such things have been said before: of the clockwork universe of Robert Boyle, and of the great syntheses of Isaac Newton and James Clerk Maxwell. More recently, in 1947, George Gamow wrote:

> We have now much sounder reasons for believing that our elementary particles are actually the basic units and cannot be subdivided further. Whereas allegedly individual atoms were known to show a great variety of rather complicated chemical, optical, and other properties, the properties of elementary particles of modern physics are extremely simple; in fact they can be compared in their simplicity to the properties of geometrical points. Also, instead of a rather large number of "indivisible atoms" of classical physics, we are now left with only three essentially different entities: nucleons, electrons and neutrinos, and in spite of the greatest desire and effort to reduce everything to its simplest form, one cannot possibly reduce something to nothing. Thus it seems that we have actually hit the

TABLE 2. High-Energy Hadron Accelerators

Machine	*Location*	*Type*	*Beam energy* (TeV)	*CM energy* (TeV)	*Status*
SPS	CERN	Fixed target	.4	.03	Operating
Tevatron II	Fermilab	Fixed target	.8	.04	Operating
ISR	CERN	pp collider	.03	.06	Discontinued
CBA					
(Isabelle)	BNL	pp collider	.4	.8	Cancelled
S$p\bar{p}$S	CERN	pp collider	.32	.64	Operating (1984/5)
Tevatron I	Fermilab	$p\bar{p}$ collider	≥.9	≥1.8	1986
Dedicated			2.0	4.0	
Collider	Fermilab	$p\bar{p}$ collider	3.0	6.0	Not recommended
UNK	USSR	$p\bar{p}$ collider	20.0	40.0	1993?
SSC	Texas	$p\bar{p}$ collider			Our dream

bottom in our search for the basic constituents from which matter is formed.

Such hubris has never survived for long. Pions were found in 1947, strange particles came soon afterwards, and the 1960s saw a virtual population explosion of new "elementary particles."

True, the standard model does explain a great deal. Nevertheless, it is not yet a proper theory, principally because it does not satisfy the physicist's naive faith in elegance and simplicity. It involves some 17 allegedly fundamental particles and the same number of arbitrary and tunable parameters, such as the fine-structure constant, the muon-electron mass ratio and the various mysterious mixing angles (Cabibbo, Weinberg, Kobayashi-Maskawa). Surely the Creator did not twiddle 17 dials on his black box before initiating the Big Bang, and its glorious sequela, mankind. Our present theory is incomplete, insufficient and inelegant, though it may be long remembered as a significant turning point. It remains for history to record whether, on the threshold of a major synthesis, we chose to turn our backs or to thrust onward. The choice is upon us with the still-hypothetical Superconducting Super Collider (SSC).

In July of 1983, the High Energy Physics Advisory Panel of the Department of Energy recommended that the highest priority be given to construction of the SSC. The recommended energy per beam of this accelerator is 20 TeV, or 20,000 GeV—this is a macroscopic energy of about 32 ergs for each proton in the beam. Head-on collisions of protons against protons will thus make 40 TeV available in the center of mass, more than 60 times the energy available at the present CERN collider and 20 times that to become available at the Fermi National Accelerator Laboratory in the near future. The committee urged, moreover, that this facility be completed and available for physics research within about a decade. The solemnity of the advice was underscored by the simultaneous recommendations that all other proposals for high-energy accelerators not be approved. This included both the Colliding Beam Accelerator, in which the Brookhaven National Laboratory had invested considerable effort, and the Dedicated Collider, a proposed expansion of the Fermilab complex.

Several arguments point to the necessity of a giant step in accelerator construction. Theorists and experimenters have settled on the parameter of greatest importance, the center-of-mass energy of 40 TeV. The High Energy Physics Advisory Panel, after long and agonizing debate, accepted this and recommended the SSC over all competing proposals. Theorists are generally agreed that new phenomena must certainly rise up and be counted at the SSC, perhaps discernible only as dim shadows at the CERN or Fermilab colliders. Indeed, experimenters fortunate enough to work at the CERN collider have already reported a handful of rare "monojets"—curious events that do not appear to be explainable in terms of the standard model. Through a tiny window, we may be seeing the new and confusing phenomena that may well require the SSC for their unraveling. Besides the arguments of experiment and theory, there is history: A great leap forward in physics occurs, regularly, at each significant jump in energy; in our projections for the 1990s, multi-TeV physics is where the new action will be.

The major parameters of the SSC, 40 TeV of center-of-mass energy and ". . . up to 10^{33} particles/cm^2 sec of luminosity," are imposed by the scientific goals. The power and the incompleteness of the standard model conspire to indicate a domain of energies where data are sure to resolve the dilemmas blocking progress in particle physics. The objective is the "1-TeV mass scale." This means that collisions should easily be capable of exploring this mass scale—for example, by producing a suspected new particle with a mass of a few TeV. The colliding protons are, however, complex objects, composed of quarks and gluons; it is the hard collisions of the pointlike constituents that are relevant. The standard model gives us the motions of the constituents that share the momentum of the colliding protons. The net effects are known in great detail and can be found in an article in *Reviews of Modern Physics* outlining cross sections for anticipated new physics processes (by E. Eichten, I. Hinchliffe, K. Lone, and C. Quigg, Vol. 56, p. 579, 1924). Because of the momentum sharing, one must divide the energy of the protons by about 10; so 40-TeV protons will permit one to explore the mass range up to 2–4 TeV.

The luminosity is a measure of the number of collisions of a particular kind that take place per second. The consensus of a very large number of workshops and seminars (as well as the paper by Estia Eichten and his collaborators) is that an energy of 40 TeV and an integrated luminosity of 10^{39} cm^{-2} will make possible a detailed exploration of the 1-TeV mass range.

Now what are we looking for? While we can list specific problems here for which we seek some resolution, we must acknowledge history's lesson that surprises have been our most frequent lot.

While we believe that a unified theory of the strong, weak and electromagnetic forces must be correct, confirming it and filling in the details require data to resolve some unanswered problems.

The major problem has to do with the symmetry of the triumphant electroweak theory—namely, that it is broken, and, in particular, that the W and Z masses are large whereas the very proper photon has zero mass. This problem has been treated by postulating the "Higgs boson," a particle whose mass and interaction properties are not specified by the symmetry. It is a study of this solution to the mass problem that points to the 1-TeV scale, below which some new phenomena, some new physics, must show up. The problem is more general: In all honesty, we have no real insight into the ultimate origins of the masses of any of the basic particles. They are simply parameters in the standard model, and so we come again to the 17-parameter problem. The 1-TeV mass scale comes in because of the close involvement of these parameters with the electroweak symmetry—the so-called Higgs sector.

Perhaps we have gone astray to assume that the quarks and leptons are really primordial. Perhaps they are composites of more basic "prequarks." If so, a new force must exist, and unification must remain incomplete until we understand it. Significant probes of the pointlike nature of electrons, muons and quarks must be at least at the 1-TeV scale, as suggested by the success of the standard model. Here we could add that we do not understand why there are two "photocopies" of the first generations (u, d, e, ν_e) and we do not know if there are additional generations.

Symmetry has been the guiding light of our world view, yet there is a long-puzzling violation of CP invariance in weak interactions. Here again, there is a connection of some kind to the Higgs boson and therefore, again, to the 1-TeV scale.

To address these and other problems, a very large number of theoretical papers have attempted to extend the standard model. Many new concepts have been introduced employing key words such as "technicolor," "hypercolor" or "supersymmetry." These often elegant ideas are unencumbered by any experimental facts. The SSC is needed to provide data that will guide physics to a true understanding.

Finally, we should note a profound novelty in the state of particle physics in the SSC era—namely, the new joining of this subject with cosmology and the data from that great accelerator-in-the-sky, the early universe. The 40-TeV machine will allow us to study matter in a state equivalent to 10^{-16} sec after the Big Bang; the data obtained may prove crucial to an understanding of how we all got here.

Perhaps we have convinved you that the "desert"—a large energy domain above the W mass containing no new physics—is a mirage. Perhaps you will concede that startling new discoveries are likely to be made with the SSC. Still, you may argue against it. Particle physics is no longer "relevant," you may say. The standard model is all we know and all we *need* to know for technology, and indeed for all of science but cosmology and elementary-particle physics itself. Our discipline, the critic continues, seems to have turned in upon itself and no longer relates to the rest of the scientific endeavors of mankind. It is a kind of art for art's sake, although far more costly to pursue. Again, the critic asks, who needs the SSC?

We respond in several ways—in terms of challenge, spinoff, pride, and duty.

CHALLENGES. Consider Arthur, an intelligent alien from a distant planet, who arrives at Washington Square (New York City) and observes two old codgers playing chess. Curious Arthur gives himself two tasks: to learn the rules of the game, and to become a grandmaster. Elementary-particle physics resembles the first task. Condensed-matter physicists, knowing full well and with absolute certainty the rules of play, are confronted with the second task. Most of modern science, including chemistry, geology, and biology since the fall of vitalism, is of the second category. It is only in particle physics and cosmology that the rules are only partly known. Both kinds of endeavor are important—one more "relevant," the other more "fundamental." Both represent immense challenges to the human intellect.

Challenge has another aspect. As physicists pursue higher energies and the ever finer structure of matter, the task becomes more difficult. The construction of great machines, elaborate detectors and powerful data-handing techniques brings us to the cutting edge of modern technology. Our workers and our factories will be compelled to confront all but insuperable technical obstacles. Meeting these challenges will make American industry better able to compete, produce and flourish in an increasingly technological society.

SPINOFF. The following arguments are well known to our colleagues in physics. They are used, and validly so, by all fields. Indeed they are strengthened by cross fertilization. High-energy physics contributes its share to the benefits that physics brings to society.

The design, construction and operation of a large accelerator in a cost-effective manner demands technological innovation that can be of considerable value elsewhere in our society. Intense study of superconducting magnets can be important to many socially relevant technologies: super-rapid transit, energy-storage systems, electrical power transmission, for example. The accelerator laboratories substantially advanced the technology, which had grown out of basic materials science. The SSC will require excavation of a very large tunnel; search for cheaper tunneling techniques will produce patents that may prove important for sewage systems, subways, and the like.

History offers many examples of past successes: Developments originating in particle physics have had an impact on computers and computer science, cryogenics, copier technology, medical diagnostics and treatment, synchrotron light sources, industrial and medical accelerators, and petroleum exploration and recovery—to name only a few examples. We should also note that the highest-priority devices in both nuclear- and materials-science programs are high-energy electron accelerators that are derived from the high-energy accelerators pioneered at Cornell and Stanford.

In addition to new and improved technologies, particle physics yields highly trained scientists accustomed to solving the unsolvable. They often go on to play vital roles in the rest of the world. Physicists trained in our discipline can be found in large numbers outside it, happily and gainfully employed, but doing something else and doing it well. One of "us" recently won the Nobel Prize in chemistry for the discovery of the genetic repressor. Another won it in medicine for the invention of the CAT scanner. Andrei Sakharov, who explained the origin of matter in the "hot Big Bang," went on to win the Nobel Peace Prize in 1975. Many of us have become important contributors in the world of energy resources, neurophysiology, arms control and disarmament, high finance, defense technology, and molecular biology. There is even an occasional artist or author.

High-energy physics continues to attract and recruit into science its share of the best and the brightest. If we were deprived of all those who began their careers with the lure and the dream of participating in this intellectual adventure, the nation would be considerably worse off than it is. Without the SSC, this is exactly what would come to pass. We acknowledge that other components of fundamental physics have equally valid claims, but let's think of the entire activity: Can we have fundamental physics without this subject and its scientifically compelling next step?

PRIDE. Although pride is one of the seven deadly sins, we are proud of the successes of our predecessors, and proud of our country, which has generously supported the study of the most fundamental structure of matter. Physics is an international discipline and has operated in a competitive-collaborative mode since Galileo. However, true collaboration requires rough equality. Yet, most of the recent discoveries in particle physics were made abroad. Gluons were first found in Germany, where the PETRA collider holds the record for e^+e^- collisional energy. CERN, after triumphantly revealing neutral currents in 1973, went on to discover the W, the Z, and the top quark, and now presents us with a bewildering array of anomalies. Many of our colleagues approach us to find out "what's up?" in particle physics. The usual answer has been, "Construction has just started on the HERA machine, a giant electron-proton collider," or, "CERN's Large Electron-Positron Collider, a 27-km ring near Geneva is underway," or, "The CERN collider has restarted brilliantly," or, "PETRA is now running at 47.6 GeV," or, "Europe will surely convert the LEP tunnel to a large hadron collider in the 1990s" or, "Perhaps we will have the SSC by 1994." More and more, American accomplishments either recede into the past perfect or dangle in the future conditional while the Europeans pursue the present indicative. Of course, as scientists, we must rejoice in the brilliant

TABLE 3. The Standard Model

The fundamental particles of the standard model are six quarks and six leptons. The quarks come in three "colors," r, y, b. The particles interact via forces, described in quantum theory as gauge-invariant fields whose quanta are bosons having spin 1. The electroweak force is carried by the photon, Z^0, W^+ and W^-. The strong (quantum chromodynamic) force is mediated by gluons. The fourth force, gravity, is not encompassed in the standard model. A fifth force, responsible for breaking the symmetry of the rest, is deleted from the tables below out of a sense of ignorance.

	1st generation	*2nd generation*	*3rd generation*
Quarks	u	c	t
	d	s	b
Leptons	e	μ	τ
	ν_e	ν_μ	ν_ν

Forces	*Carried by*
Electromagnetic	Photons
Weak	W^+, W^-, Z^0
Strong	Gluons

achievements of our colleagues overseas. Our concern is that if we forgo the opportunity that SSC offers for the 1990s, the loss will not only be to our science but also to the broader issue of national pride and technological self-confidence. When we were children, America did most things best. So it should again.

A SENSE OF DUTY. This motivation for the SSC is the most difficult to explain, but it is the driving force of the particle physicist. Faith in the underlying simplicity of nature—quite unjustified, to be sure—has time and again led to discovery. We are amazed at the incredible successes of twentieth-century science, and at its enormously positive (and, regrettably, sometimes negative) effect upon everyday life. The universe astonishes us by its very comprehensibility. In this we find our call: Being born upon an obscure planet located at the rim of a middling galaxy among a hundred billion galaxies of an aging universe, it is our sacred duty to know its deepest secrets, as well as we are able. Dolphins and chimpanzees can be made to speak, after a fashion. Yet, only humans will look at the stars with wonder and find it necessary to understand just what they are and how they work and why we are here to see them.

No better mousetrap or wrist TV here—just the triumph of human imagination. It is simply *the need to know* that compels us to build a bigger and better accelerator and to approach an understanding of the mother of us all—the Big Bang—and its curious by-product, the matter of which we are made.

Colleagues are careful to insist that particle physics at the high-energy frontier can have no direct effect upon our technology. If this turns out to be true, it will be the first time in the history of science. But what can we say with certainty about technologies of the future? Human society, aided by science, should be better able to cope with the vicissitudes of life on Earth than the dinosaurs could. Yet, they flourished for 300 million years. One may hope that our species may do at least as well. We cannot argue reasonably that the pure physics of today will not become essential to the technology of a distant tomorrow.

As now conceived, SSC is a double ring of superconducting magnets in an underground tunnel 90–180 km in circumference (see Figure 1). It will have four or six interaction regions where counter-rotating protons make head-on collisions to generate 40 TeV in the center of mass. In January 1984, about 150 physicists and engineers from throughout the U.S. gathered at the Lawrence Berkeley Laboratory and, by May of the same year produced the "Reference Design Study." This document was intended to narrow the uncertainties of costs and schedule and to determine the most fruitful topics to be addressed by research and development (R&D). The reference design breaks no dramatic new ground in fundamental accelerator science. It is unabashedly a scaled-up and improved version of Fermilab's superconducting Tevatron, making use of both the experience developed there and the intensive R&D in magnet technology now taking place at Brookhaven, Fermilab, LBL, and the Texas Accelerator Center. Other systems that will be improved are cryogenics and computer controls, both of which will be applied on an unprecedented scale. The enormous size of the SSC is illustrated in Figure 2.

Work on superconducting magnets began in accelerator labs after the discovery of hard superconducting materials, in about 1960. Soon, very large magnets were built for spectrometers and other instruments. In the late 1960s, laboratories started to develop pulsed superconducting magnets, adding the complication of rapidly varying fields. At Fermilb, R&D

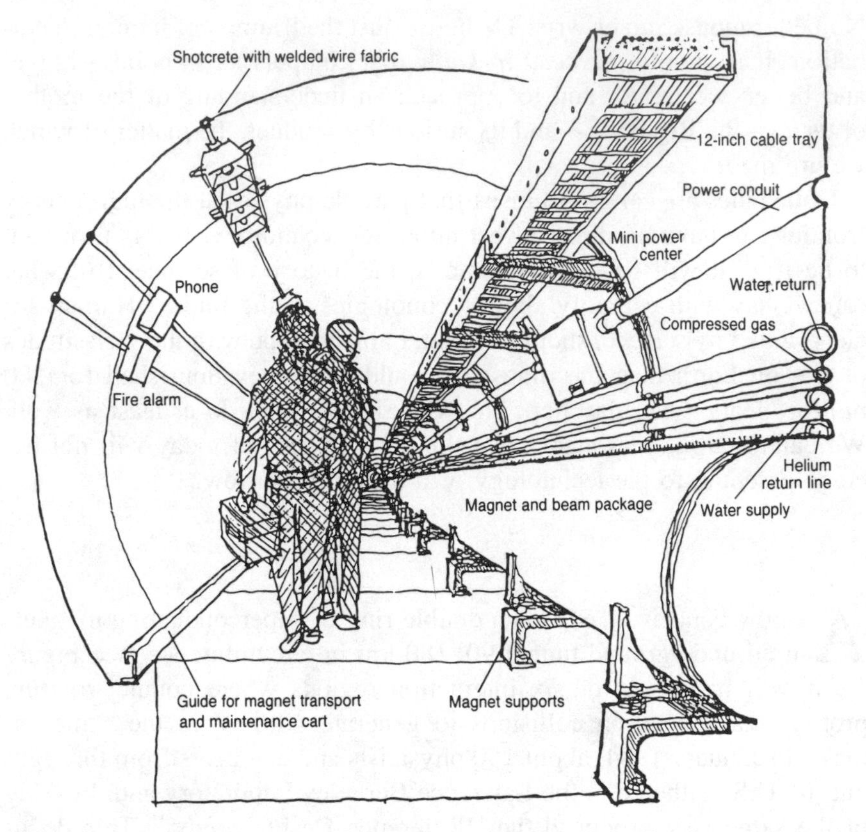

FIGURE 1. *Perspective view of the main ring tunnel in the Superconducting Super Collider design.*

work began in 1972–1973 and benefited from the experience of many laboratories, most notably the Rutherford Laboratory in England, and LBL and Brookhaven in the U.S. The construction of a ring of superconducting magnets began in July 1979. The entire 6-km-circumference ring of over 1,000 magnets was cooled to 4.5 K in May 1983. In July 1983, acceleration of protons was achieved. Experiments were started in October 1984 and today 800-GeV beams are being delivered routinely to eight targets. There is every indication that a superconducting machine will be a reliable and efficient device.

The Tevatron has raised the available energy from 400 to 800 GeV (the goal is 1,000 GeV) and, at the same time, has reduced electrical power consumption by a factor of four. The basis of confidence in the feasibility of an SSC rests on the success of the Tevatron accelerator.

FIGURE 2. *The scale of the Superconducting Super Collider (if 5-T magnets are used) is shown in comparison with the Tevatron at Fermilab (the smallest ring) and LEP at CERN (the somewhat larger ring). All three rings are superimposed to scale on the environs of Washington, DC. Note that the SSC is about the size of the Washington Beltway. This NASA photograph was made in November 1982 from an orbiting Landsat.*

The reference design envisions an accelerator complex involving a " . . . central office–laboratory building, sized to accommodate 3,000 full-time and visiting scientists and staff, injector facilities consisting of a 200-m linear accelerator, a 1.2-km low-energy booster and a 6-km high-energy booster, designed to provide a 1-TeV beam of protons suitable for injecting into the SSC main ring." The two concentric rings of superconducting magnets provide the place where counter-rotating protons would be accelerated (each to 20 TeV), stored, and then brought into collision in the interaction regions. Typically, storage times in excess of 10 hours are expected.

Two magnet strengths are being considered in the current R&D program: a "low-field" 3-T magnet, which would imply a ring circumference of 180 km, and a "high-field" 6-T magnet, requiring at tunnel of 90-km circumference (see Figures 3 and 4). The two magnets also embody different philosophies: The 3-T design depends heavily on the shaping of iron pole pieces to define the quality of magnetic field that guides the protons; in the 6-T design, the current-carrying superconducting wires determine the field shape, as in the Tevatron.

The reference design estimates a need for thousands of magnets of the bending and (quadrupole) focusing types, and for a massive cryogenic system to bring and hold these at liquid-helium temperatures. Managing a system of this scale will require an instrumentation and control system of impressive proportions.

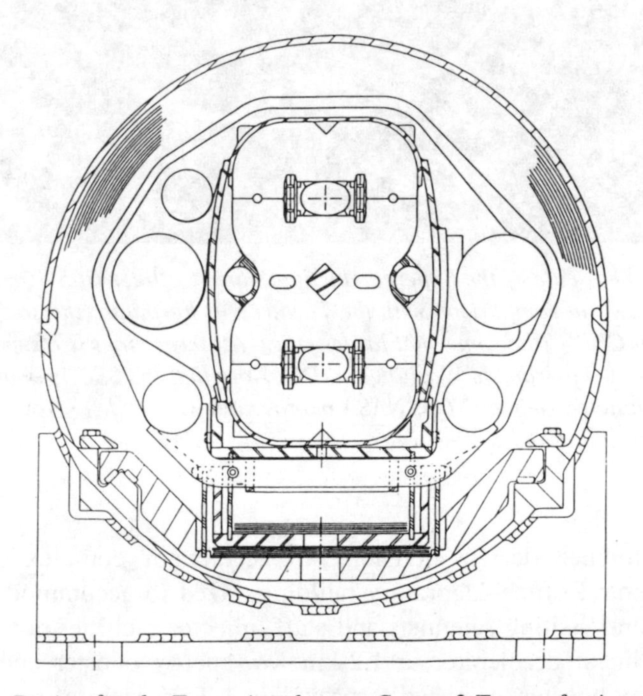

FIGURE 3. *Design for the Texas Accelerator Center 3-T superferric magnet. The magnet has two beam channels that are magnetically independent. Eight turns of 10 kiloamps are used to drive the field in each channel. The 3-T magnet was called design "C" in last year's Reference Design Study.*

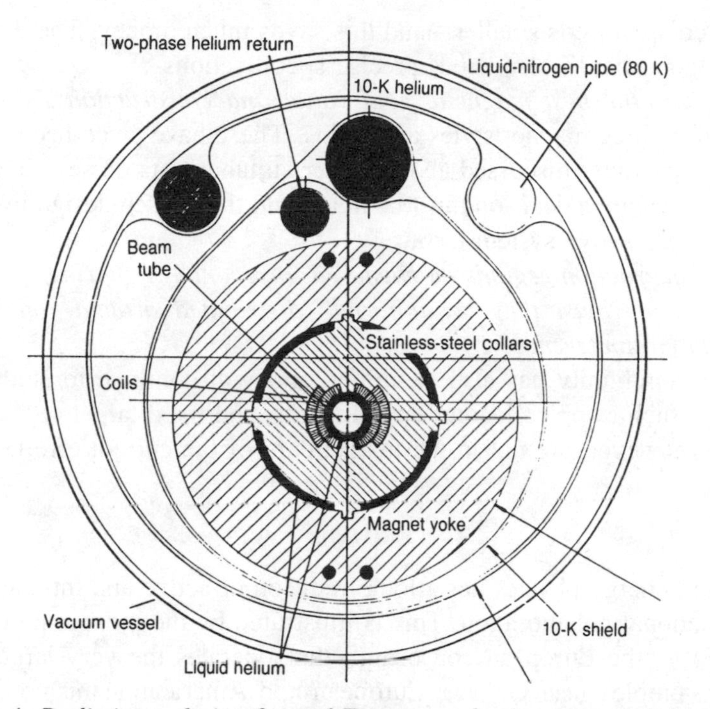

FIGURE 4. *Preliminary design for a 6-T superconducting magnet, being developed by a Brookhaven, Fermilab, Lawrence Berkeley Lab collaboration. This design, denoted "D," is a compromise between design "A" for a 6.5-T magnet and design "B" for a 5-T magnet, both of which were considered in last year's Reference Design Study. The vacuum vessel in this drawing has a 53-cm diameter.*

The Reference Design Study (RDS) and subsequent workshops have to date uncovered no obstacles to our plan to construct such a machine. What provides the unprecedented technological hurdle is the scale of the project. The major challenge is in technology and engineering, although some physics issues do remain. What is needed, the experts concluded, is about three years of hard work to design options and to invent and test cost-saving and reliability-enhancing ideas. The objective is to have the machine ready for physics experiments by about 1994.

Some of the uncertainties that must be removed before the definitive design is made center about the following questions:

What is the minimum magnet aperture that will work? The magnet aperture of the Tevatron is about 7 cm, which yields a "good-field" region of about 4 cm. With a higher-energy injector to the SSC, the required aperture is smaller—and this saves much money. The definition of a ''good field'' also needs precise specifications.

What is the best magnetic field value and construction style? RDS looked at three magnet styles and fields. These have since been reduced to the two styles discussed above. After suitable tests on several models of each design, a decision must be made and the chosen design improved during exhaustive system testing.

Can interaction regions be clustered for economic sharing of support facilities, or must they be uniformly distributed around roughly the 100-km circumference?

The community has also examined and will continue to study other options such as $p\bar{p}$ collisions, ep collisions, and fixed-target applications. These, however, are not in the mainstream of the current effort.

High-energy physics has a long tradition of active and intimate international collaboration. This is illustrated by the pioneering creation of CERN, the European consortium that operates the very large accelerator complex near Geneva. European and American teams continue to use one another's facilities easily and frequently. We have formal exchange agreements with Japan, China and the Soviet Union..

In 1975, high-energy physicists organized the International Committee on Future Accelerators. ICFA had two missions: to use its best efforts to enhance communications and minimize duplication of frontier facilities, and to look ahead to the time when the resources required for the next energy level would require worldwide collaboration. Recently, this grass-roots movement was supplemented by promulgations coming down from the Economic Summit. Here the leaders of the industrial nations (European Economic Community, Japan, U.S., Canada) have selected a number of scientific and technical fields and have in effect committed themselves to the intellectual prosperity of these fields, while urging that a coherent, collaborative, long-range plan be presented. The London Summit of June 1984 was followed by a high-energy physics meeting in Brussels in July 1984. There, a committee was established to examine the problems of international collaboration in the long-range planning of new facilities. Recognizing this task to be a two-to-three-year process, the committee requested an interim report by June 1985.

The U.S. community, moderately optimistic about SSC as a result of the decision of the secretary of energy (in August 1984) to proceed with the R&D plan, nevertheless took the summit message very seriously. Our European colleagues are in the midst of an ambitious program of accelerator construction at CERN and DESY (in Hamburg). When this program is completed in 1990, they will have expended close to $2 billion on capital facilities, including detectors. (Here we have tried to do the accounting on the U.S. system where, for example, salaries are included in the cost estimations). After that, it is conceivable that Western Europe could join with other countries to make contributions to the SSC. Some of the candidate nations that might assist by accepting construction responsibilities are Japan and Canada. Other modes of collaboration are also possible. Obviously there are risks and, not so obviously, vast sums of money are unlikely to be saved by the host country. Nevertheless, as facilities escalate in cost, both financial and intellectual, there is widespread recognition that a coherent international plan makes sense. We look to the Sherpas to guide the summiteers toward such a plan over the next few years.

The cost of the SSC was carefully estimated by RDS at about $3 billion (1984 dollars). This estimate includes a contingency fund, suggested by the Department of Energy, of about 20 percent. One must add to this the cost of detectors ($4–8 billion) and preoperations ($1–2 billion).

Much has been learned from CERN's 600-GeV (center-of-mass) collider about the required properties of a general-purpose detector. The Tevatron Collider Detector Facility at Fermilab will cost $60 million. Estimates for scaling this 2-TeV detector to 40 TeV run as high as $200 million. Special-purpose detectors designed for more specific researches are usually far less expensive.

The track record of high-energy physics construction projects is very good. In view of the essentially conservative technology and in view of the now characteristically exhaustive scrutiny of cost estimates for major projects, it is extremely unlikely that overruns will be incurred. On the other hand, two to three years of R&D could produce significant savings. International collaboration may further serve to reduce the cost to the U.S. taxpayer. The largest previous high-energy-physics construction, Fermilab, cost about $900 million (in 1984 dollars) on the same basis. Thus we are facing an almost fourfold increase for a facility of the 1990s over the amount spent for the accelerator of the 1970s.

In view of the universal increase in demand for more sophisticated equipment, this increase is not out of line. It's just a very large sum of money. After construction, and assuming an eventual constriction of the funding for the rest of high-energy physics, the budget for high-energy physics can return to its pre-SSC level. The annual operating budget of the new facility has been estimated at about $200 million. To this sum, one would have to add about $50 million for detector and machine improvements. High-energy physics in 1985 is supported at about $600 million. Clearly, the new machine will draw major effort and resources from the ongoing program. There doesn't seem to be any reason why the U.S. should not be able to afford this and other research facilities. It is crucial, of course, that these activities pass stringent tests to determine their scientific value.

There is concern that so large a project will adversely affect other deserving physics programs. A study of the funding history over the past 25 years does not support this concern, as Figure 5 shows. Although we cannot predict the future pace of science funding, it is likely that the public and its representatives in government will continue to appreciate

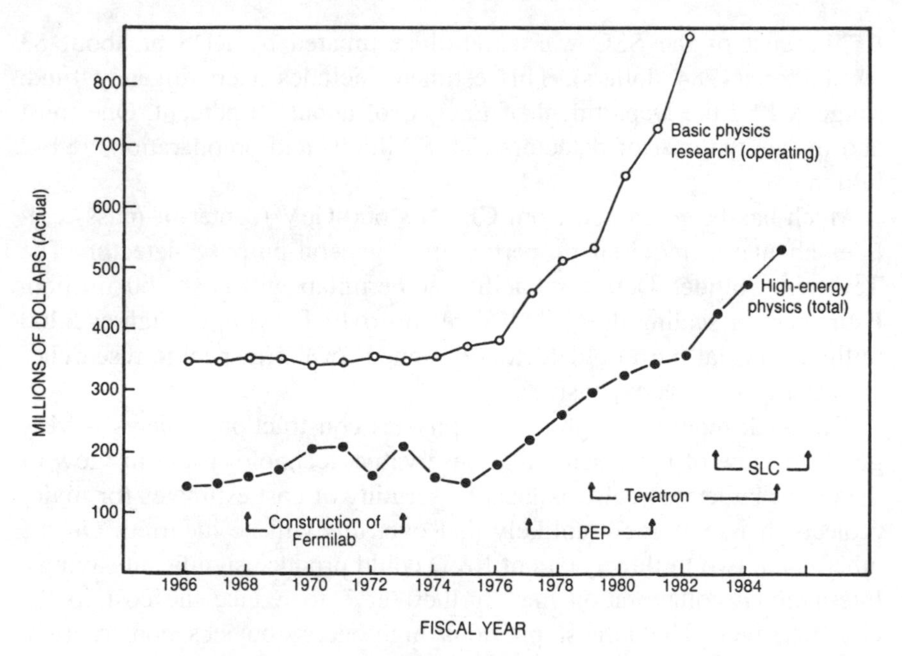

FIGURE 5. *Federal budgets for physics. The vertical axis is in actual dollars, uncorrected for inflation. (Source: Physics Survey, 1984)*

good science as a necessarily increasing proportion of the federal budget. But this esteem will not persist if the scientific community loses its vision and exuberance and becomes fretful and divisive. Yes, we have national budget deficits, and yes, we have urgent social problems. But if we have faith in the enduring future of the nation, then basic research must go on with reasonable stability, and at a pace that is perceived to be viable by the young scientist.

The high-energy-physics community has been reasonably responsible in recommending the termination of older, but still scientifically useful, facilities to provide funding for accelerators that could address more crucial issues.

At each stage, we were compelled by increasing cost and complexity to assemble larger teams and to stretch experiments over longer durations—the idyll of a backyard accelerator is now for very few, and commuting across the country is the norm. No one likes this complication, and it does require enormous attention to ensure that we are still attending to our students, our junior faculty, our replacements. For example, until recently, universities have been the training grounds for accelerator builders and facility directors. As accelerator technology has become much more sophisticated, individuals have become specialized as accelerator physicists and particle physicists. It is difficult to generate accelerator leaders and experts, as the number of accelerators decreases and they become divorced from the universities. The SSC is one more step in this process, but we see no alternative that preserves the scientific vitality, no, the *validity* of the activity. It is our opinion that high-energy physics must go in this direction or terminate the 3,000-year-old quest for a comprehension of the architecture of the subnuclear world.

Modern particle physics has a rich heritage, following a track—consciously or not—set out for us by the ancient philosophers. This enduring quest has produced several great intellectual revolutions, two in this century: one in the perception of space time and causation (relativity), the other in the understanding of the behavior of matter at the atomic level (quantum theory). We must also note the gradual establishment of a scientific basis of technology and the interdependence of sciences through the overlapping content and the spread of instruments of science. Particle physics has, as its intellectual neighbors, on one side cosmology, on the other nuclear physics. The techniques and devices evolved in the accelerator laboratories have found ready applications in other fields of science. Symmetrical benefits have been received, both intellectual and technological, from all the subdisciplines. However, the thing we all share, above all else, is the sense of wonder and awe at the distance we have covered toward comprehension of our universe.

We are now asking our fellow physicists to join us, however vicariously, in a very great adventure: nothing less than a giant step in the continuation of our collective ambition to strive for a deeper understanding of that nature within which humankind is embedded.

Passing the Torch

Americans win the lion's share of Nobel prizes in physics, chemistry, and medicine. This is frequently, but wrongly, quoted as evidence of the health of American science. In fact, it is interest earned by past investments. Prize-winning research was done 10 to 25 years ago, when research was more generously supported than it is today. Prize-winning scientists received their crucial precollege education in the first half of the twentieth century. Since then, European investment in science education and research has been much larger, per capita, than ours. This will become apparent with the Nobel prizes of the next two decades.

This country was once the unquestioned technological hub of the world. Today, most of our industry is in deep trouble. Steel, ships, sewing machines, stereos, and shoes are lost industries. Japanese cars are generally thought to be cheaper and better made. Proud RCA has become a distributor of Japanese goods, assembled in Korea. American Motors is controlled by the French government. We even buy Polish robots. Advanced electronics and computers are soon to be challenged by the Japanese. As we exhaust our heritage of capital and raw materials, Americans will no longer be able to afford the technological society to which they have become accustomed. We shall be left with our Big Macs, our TV dinners, and, perhaps, our federally subsidized weapons industries.

How is it that the forces of the marketplace have failed us? Is it too late for a technoloical renaissance? We have been leading our young people away from science and technology. The Vietnam experience, the failures of our nuclear power industry, and the threat of nuclear holocaust are partly responsible. The almost complete lack of precollege teachers with competence in science and math has played a role. The forces of the marketplace have driven the few good ones into the desperate, but better paying, arms of industry. Who will our industries turn to next? Most of our high school students do not understand algebra or chemistry. We cannot count on them to reconstruct our technological society.

I was educated in the public-school system of New York City. The State Regents exams demanded serious, substantive, and standardized curricula. Reading and math levels were tested often, and students were assigned in accordance with their skills. This sort of testing is unpopular today. Students are put into "open classrooms" and told to "do their thing." Self-expression is important; grammar and history, let alone science and math, are not. Our schools are fascinated by complicated and expensive scientific toys and "audio-visual aids." What they really need are scientifically literate teachers. Frogs, cow hearts, scalpels, siphons, a few leaves, pond water, the night sky, an inexpensive microscope, a good chemistry set, and some batteries, wires, and bulbs are enough to teach a lot of science. What my kids get is prepackaged commercial pseudo-educational pap like "magic powders." But they cannot tell an oak tree by its leaf. Five hundred nonscientific Harvard undergraduates take my core course, "From Alchemy to Quarks." Many of them cannot name one chemical element, or identify one planet or constellation. They will become famous sociologists or political scientists, but they are violently allergic to numbers. They suffer from "dysmetria," as do most Americans. Perhaps they are the people who will be entrusted with the U.S. budget a few years hence.

My father saw Halley's comet in 1910. Then he witnessed the explosive technological growth of this country, of which he was very proud. He explained to me when I was a child that Halley's comet would return in 1985, and that American scientists would voyage into space to meet it and solve its mysteries. He would not have been pleased to know that it will be the Russians, the French, and the Japanese who will launch the cometary probes. We could do it, of course, and do it best, but we have chosen not to. The torch of scientific endeavor has been passed to other peoples.

We were once the leaders in high-energy physics, my own specialty. We invented atom smashers, and, until recently, we had the biggest and the best. Since the opening of the Intersecting Storage Rings at CERN in 1971 and the CERN Collider in 1981, we have been completely outclassed by our Western European friends. The exciting field of electron-positron collisions was pioneered in France, Italy, and the Soviet Union in the 1960s, but we were triumphantly supreme in the 1970s. However, since 1978, we have been beaten by the Germans. Because Western Europe spends more than twice as much on high-energy physics as the U.S., the future of this field in our country is not very rosy. Ironically, it is the force of the marketplace that impels the Europeans. Perhaps more clearly than we do, they see technology as the key to a healthy industrial society.

Teaching the Lowest Common Denominator

My wife, Joan, ran for the Brookline, Massachusetts, school committee on a platform of "More Math and Science." She lost to the forces of "Equity not Excellence," or "Lesson Less for Moron More!"

We Americans aim at the lowest common denominator, and we aim low. Our children's training in math is scandalous. Our best are not nearly good enough, and most high-school graduates suffer from Acquired Incurable Dysmetria, otherwise known as Fear of Mathematics.

Recent studies establish beyond doubt the relative mathematical illiteracy from first grade onward of American vs. foreign school children. The February 14, 1986 issue of *Science* reports on a comparison of 2,300 pupils from Taiwan, Japan and Minneapolis. As early as first grade, our students have fallen behind their Asian counterparts, and by fifth grade, the effect is unmistakable.

A more ambitious study, sponsored by the U.S. Department of Education, involving 6,648 American children as well as students from many other nations, reveals that the disease progresses relentlessly with age. Our eighth graders ranked thirteenth among the 17 countries studied. Japanese, Dutch and Hungarian students outclassed our children by as much as ours ranked above the Nigerians and Swazilanders.

Ours is a technological society needing mathematically competent workers to staff its robots and computers, and realize its high-tech hopes. Another Education Department survey focuses upon the top 5 percent of students from 12 countries. This time, America comes out dead last. Our very best students, in their senior year of high school, are no better (mathematically) than average students from Finland, and quite inferior to randomly chosen Japanese students.

College freshmen, by and large, cannot pursue the professions of their choice. Mathematical literacy is essential for the study of physics, chemistry, mathematics, computer science, engineering, economics, and for premedical or predental programs. How many of our youngsters have been disqualified by the weakness of their precollege education? Most of them.

America has dozens of first-rate schools for postgraduate study in science and engineering, but not nearly enough qualified American applicants. More than half of our graduate students are foreigners, and the ratio is increasing. Many stay in this country after their studies to take high-paying high-tech jobs, while undereducated Americans sweep their floors and cut their grass.

Why are American students so bad at math? According to *Science,*

> Most American mothers interviewed in this study did not appear to be dissatisfied with their children's schools. . . . The children, faced with parents who generally are satisfied and approving of what happens in school, must see little need to spend more time and effort on their schoolwork.
>
> The poor performance of American children in mathematics thus reflects a general failure to perceive that American elementary school children are performing ineffectively and that there is a need for improvement and change if the United States is to remain competitive with other countries. . . .

There are other reasons: Few American teachers understand the math that they are supposed to teach. If they did, they could (and would) earn far more in industry. Textbooks in science and math are often written with explanations that make no sense and problems that cannot be done. As a parent of four children in the public schools, and as a practicing scientist, I know.

Then there is television. Is there an American child who spends as much time on homework as he or she does glued to the boob-tube? Not likely. Learning mathematics, like learning the oboe, takes time and practice.

Will things change? Not likely. Despite the warnings sounded in the 1983 presidential commission report, *A Nation at Risk,* the situation has become far more alarming. One by one, our industries fall victim to superior foreign technologies: steel, shoes, stereos, ships, silicon chips, cars, robots, textile and agricultural machinery, and soon, computers and airplanes. No wonder there's a whopping trade deficit. It's not the cheap labor nor the strong dollar so much as it is the lack of imaginative engineering or what we once proudly called "know-how."

Our bright children, excluded by our criminally poor schools from doing anything useful for American technology, are forced into such professions as law and advertising. They will do to you what they did to AT&T and convince you that you like it.

Meanwhile, the French launch satellites with no loss of life using their unmanned Ariane rockets, the Soviets send us close-up pictures of Halley's comet, Canadians sell us electricity produced by safe and reliable reactors, Western Europeans find all the elementary particles, and Japanese offer cameras that even Americans can operate. While we can afford the free ride, it's not a bad deal. But watch out! The yellow brick road of downward mobility doesn't lead to the land of Oz.

Science and Violence

Knowingly or not, scientists are deeply and necessarily involved in the political and military history of mankind. The great Archimedes developed weapons of war to defend Syracuse against the Romans, but the war was lost and Archimedes was killed. Georg Brandt discovered phosphorus in Hamburg. Three hundred years later his city was destroyed by the incendiary bombs his discovery made possible. Antoine-Laurent Lavoisier—first to understand the gift of Prometheus—was executed in the aftermath of the French revolution. "France has no need for savants," said the judge. Henry Mosely—discoverer of the concept of atomic number—was allowed to fight and die in the trenches. England, too, had no need for savants. Germany's racial policies put an end to her preeminence in physics. Germany had no need for savants. And, the United States of America threw Count Rumford out as a British spy and shamed Robert Oppenheimer.

Scientists gave the world the nuclear weapon: Otto Hahn, by his discovery of fission; Enrico Fermi, by the construction of a nuclear reactor; Albert Einstein, by his famous equation and equally famous letter; Richard Garwin and Edward Teller by their contributions to the development of the hydrogen bomb; and many, many more. Like an automobile manufacturer, or a toy maker, the scientist is responsible for his discoveries.

Over the years of the arms race, the world inventory of nuclear weaponry has grown beyond all rational proportion, and it continues to grow. If it were to be used, a disaster, such as the world has rarely seen, would take place. Luis Alvarez argues that a comparable disaster—an encounter with a sizable meteorite—led to the great Cretaceous extinction 70 million years ago. So ended the 300 million year reign of the dinosaurs.

Our civilization is only a few thousand years old, yet we already faced with a similar calamity, and one of our own making.

My view is simple and naive. I believe in the eventual and complete elimination of nuclear weapons: absolute nuclear disarmament, or AND, a conjunction which suggests a possible future to our civilization.

Many of those present prefer the concept of mutual assured destruction, or MAD. Mad as it is, it *has* been responsible for the avoidance of nuclear war for a generation. Nonetheless, I find the concept repugnant, and unacceptable as a permanent crutch for society. MAD has led to a continuing arms race, which is not merely financially delibitating and morally offensive, but it is dangerous: So dangerous as to make an eventual nuclear holocaust inevitable. MAD is an effective short-term remedy with lethal long-term consequences. There are all too many plausible scenarios which could lead to nuclear war: Accident, miscalculation, irresponsible and misinformed leaders, unauthorized access, undue provocation, *ad nauseum.* Who can guarantee, for example, that future world leaders will have the intelligence and integrity of a Nixon or a Krushchev or a Reagan or a Brezhnev? Who would deny that any of these people, should they have chosen to do so, could have precipated a nuclear war. Kennedy came close to doing just that. Can anyone be trusted that much?

AND is impossible, you say. A certain number of small weapons can always be concealed, especially in a closed society. Perhaps this is true. Perhaps all we can do is approach AND. Let us reduce the two great stockpiles of strategic nuclear weapons. *They* need thousands of weapons because *we* have thousands of weapons. Let both sides reduce their weaponry by a factor of 10. Then, *they* will have hundreds of invulnerable missiles and so will *we*. This is still MAD, but it is 10 times closer to being AND as well. And if that works, perhaps another factor of 10 . . .

Must mutual assured destruction involve the potential destruction of entire nations? Would it not be sufficient to hold Moscow and New York hostage, and leave the rest of us secure?

The prevention of nuclear war should become the acknowledged central goal of humanity. There are many other important problems to address: poverty, hunger, disease, hatred, racism, pollution, human dignity, to name a few. But, the prevention of nuclear war comes first and foremost.

Wars have been fought throughout history. They are being fought today. They will be fought in the future. These must not be nuclear wars. Tactical nuclear weapons and theater nuclear weapons are addressed to "limited nuclear war." I do not think there can be such a war. I trust a general far less than I do a president. It is not at all clear that a treaty banning such weapons is impossible. The citizens of the world want, need, and must demand such a treaty. I prefer to counter 20,000 Russian tanks with 20,000 cheaper antitank weapons than nuclear weapons or "neutron bombs." Let's ban the little bombs first.

How much is it necessary to distort and destroy Western civilization to provide security in the ongoing arms race? Will we have evacuation drills of New York City to see whether its citizens can be salvaged? Shall we contaminate the American dream with a bomb shelter and a machine gun in each cellar? Shall we sacrifice a state or two for an MX missile site? Would you rather be a rat or red? Or, is there a more sensible way?

Nuclear weapons are about as necessary to humanity as smallpox. And, I am still fool enough to believe that we can deal with nuclear weapons in exactly the same way.

Acknowledgments

Grateful acknowledgment is made to the following publications for permission to reprint previously published material.

ELEMENTARY-PARTICLE PHYSICS AND ME is based on a talk delivered at the Centennial Celebration of the Jefferson Physics Laboratory at Harvard University, May 4, 1984. It was also the 30th anniversary of Glashow's affiliation with Harvard. The anomalous results at CERN which were discussed have since disappeared.

INTERNAL EXILE IN CALIFORNIA is published for the first time in this volume. The poem "To Abalone Unbound" first appeared in *Physics Today,* December 1984, page 19, and deplores a flood of irreproducible results. The cold fusion debacle shows that things never change in the land of the setting sun.

THE MYSTERIES OF MATTER, and interview between Sheldon Glashow and Peter Costa, was broadcast nationwide in 1989 on the UPI Radio Network as part of the public affairs series "Harvard Newsmakers." The version here is adapted from "A Conversation with Sheldon Glashow" in the *Harvard Gazette,* April 7, 1989.

A PEEK AT THE UNIVERSE was prepared as supplementary reading for the Harvard Core Curriculum course *From Alchemy to Quarks* taught by Glashow for more than a decade.

LIFE ON LOG TIME was prepared as a supplement for the Harvard Core Curriculum course *From Alchemy to Quarks.*

THE NUMBER GAME is published here in English for the first time. Along with several other contributions to this volume, it was originally solicited by Professor Antonino Zichichi for the weekly feature *Cultura Moderna* of the newspaper *Il Tempo,* where it appeared in an Italian version on March 3, 1981.

WELCOME TO UBS was prepared as a supplement for the Harvard Core Curriculum course *From Alchemy to Quarks.*

ARE WE ALONE IN THE UNIVERSE? was prepared as a supplement for the Harvard Core Curriculum course *From Alchemy to Quarks.*

THE BIG PICTURE was prepared as a supplement for the Harvard Core Curriculum course *From Alchemy to Quarks.*

WHAT IS AN ELEMENTARY PARTICLE? was delivered on November 30, 1984 to an assembly of high-school students at the University of Chicago as part of The Illinois Science Lecture Association Christmas Lectures. The manuscript was prepared by the organizers from a tape recording but, to the best of the author's knowledge, was never before published.

THE HUNTING OF THE QUARK was originally published in *The New York Times Magazine,* July 18, 1976. Copyright 1976 by The New York Times Company. Reprinted by permission.

QUARKS WITH COLOR AND FLAVOR is adapted from an essay originally published in *Scientific American,* October 1975, pages 35–50. Copyright 1975 by Scientific American, Inc. All rights reserved.

THE INVENTION AND DISCOVERY OF THE CHARMED QUARK appeared in a revised version in *Il Tempo* on September 24, 1983. It is published here in English for the first time.

ANTINEUTRONS AND GEOLOGY appeared in a revised version in *Il Tempo* on January 28, 1984. It is published here in English for the first time.

ELEMENTARY-PARTICLE PHYSICS AS A WASTE OF TIME AND MONEY is published for the first time in this volume.

DOES ELEMENTARY-PARTICLE PHYSICS HAVE A FUTURE? was originally published in *The Lesson of Quantum Theory,* edited by J. de Boer, E. Dal, and O. Ulfbeck, Elsevier Science Publishers B.V., 1986, pages 143–153. It is adapted from a talk presented at the University of Copenhagen on October 4, 1985 for the Niels Bohr Centenary Symposium. Twenty-five years before, at the completion of his post-doctoral stay at the Bohr Institute, Glashow had submitted his paper outlining the algebraic structure of the electroweak theory.

BIG THINGS, LITTLE THINGS was originally published as ''Closing the Circle'' in the October 1989 issue of *Discover*, pages 66–72.

TOWARDS A UNIFIED THEORY: THREADS IN A TAPESTRY is adapted from a talk presented in Stockholm on December 8, 1979 upon acceptance of the Nobel Prize in Physics. Copyright 1980 by The Nobel Foundation.

UNIFIED THEORY OF ELEMENTARY-PARTICLE FORCES, by Sheldon L. Glashow and Howard Georgi, was originally published in *Physics Today,* September 1980.

GRAND UNIFICATION was written at the 1980 Scottish Summer School at St. Andrew and was originally published in *New Scientist,* September 18, 1980.

ON THE WAY TO A UNIFIED FIELD THEORY was originally published in *Progress in Scientific Culture,* Ettore Majorana Center for Science and Culture, Erice, Italy, 1982.

TANGLED IN SUPERSTRING appeared in a revised version in *Interactions: A Journey Through the Mind of a Particle Physicist,* by Sheldon Glashow and Ben Bova, Warner Books, 1988, pages 22–25.

GRAND UNIFICATION AND THE FUTURE OF PHYSICS is adapted from an essay prepared for publication in *Il Tempo.* It is published here in English for the first time.

SSC: MACHINE FOR THE NINETIES, by Sheldon L. Glashow and Leon M. Lederman, was originally published in *Physics Today,* March 1985, pages 2–11. It led to a spirited exchange of letters to the editor of that journal.

PASSING THE TORCH was originally published in *Physics Today,* April 1983.

TEACHING THE LOWEST COMMON DENOMINATOR was originally published as "Lowest-Common-Denominator Math" in *The Los Angeles Times,* March 30 1986, part IV, page 5. It reflects Glashow's growing concern with American science and math education.

SCIENCE AND VIOLENCE is adapted from the opening talk at the *International Seminar on the World-Wide Implications of a Nuclear War* presented in Erice on August 14, 1981. It appears in the Proceedings of that seminar edited by E. Etim and S. Stipcich, Frascati, Italy, pages 7–10, 1982.

The author is also grateful to the following people for permission to reprint coauthored material.
Peter Costa, director of news and public affairs at Harvard University.
Howard Georgi, professor of physics at Harvard University.
Leon M. Lederman, Frank L. Sulzberger professor at the University of Chicago.

FROM THE SERIES EDITOR

It has been a great privilege to have been part of Masters of Modern Physics. Working with the physicists whose essays have been selected for this series revealed once again that those who display the character and patience to accomplish great things in science need not elevate themselves above the rest of us. Their modesty, dignity, and thoughtfulness was exhibited throughout. My deep appreciation goes first to the authors who graciously permitted their contributions to be published here and who worked carefully and patiently to see these essays appear in print.

My gratitude also goes to the distinguished members of the series advisory board—Dale Corson, Samuel Devons, Sidney Drell, Herman Feshbach, Marvin Goldberger, Wolfgang Panofsky, and William Press. Their intellectual breadth and continuing advice have helped immeasurably in bringing out these volumes.

From the start, the series received unhesitating encouragement from the American Institute of Physics' Books Subcommittee. My thanks especially to Robert Beyer of Brown University, who chaired the group earlier, and to Gerald Holton of Harvard University, the present chair. The AIP is fortunate to have appointed such eminent scholars. Best of all, they have been gracious, literate, and witty.

My thanks also go to those on the staff of the American Institute of Physics who helped in large ways and small. I am especially grateful to Kenneth Ford, executive director, whose enthusiasm assured support at the highest levels. Dr. Ford also participated as a scout, travelling to the USSR to persuade leading Soviet authors to include their work in the series. My appreciation also goes to Publishing Director Robert Baensch for championing the project and for assuring it receives wide attention. My thanks, too, to AIP Books Manager Tim Taylor for orchestrating events so skillfully in the final days of publication, and to the entire AIP production staff, including in particular Larry Feinberg, Doreene Berger, Donna Colaianni, Andrew Prince, and Christa Turley. I also wish to thank Robert Marks, now at the American Chemical Society, whose early encouragement made the series possible.

Without the keenly sensitive and highly professional members of my staff—all of whom played a part in this project at one time or another—these books would never have appeared. Elaine Cacciarelli, senior associate at Robert Ubell Associates, never hesitated to lend her hand nor stinted in offering her intelligence on important questions or crucial details. Managing Editor Barbara Sullivan shepherded these books along smooth and rocky paths. She deserves special credit for her steadfastness, skill, and lively wit—and especially for her ability to work closely with the authors and the publisher's staff. Mark P. Meade and Michelle Levy-Leavitt were equally resourceful.

Large and continuing projects such as these require the help of many. Some bring intellectual gifts; others contribute their editorial skills; still others, warmth and understanding. It is impossible to measure Rosalyn Deutsche's contribution, since she has made herself indispensable in every way.

Robert N. Ubell

Index

accelerators. *See also* Superconducting Super Collider, 172, 210–211, 268 (Table 2)
Adams, John Couch, 35
Adler, Stephen, 150
aleph-null, 55–58
aleph-one, 58
algebraic numbers, 57–58
alpha decay, 101
alpha particle scattering, 123
aluminium, 113
Alvarez, Luis, 11, 291
amino acids, formation of in primordial Earth atmosphere, 81
Andromeda galaxy, 37, 38 (Fig. 3)
antimatter, 96, 123, 130
antineutrinos, 136, 168–170
antiprotons, 124, 130, 143
Apollo, 35
Applequist, Thomas, '156
Aristotle, 30
Arp 220 (starburst galaxy), 78
astrology, 27
astronomy, 48
asymptotic freedom. *See also* quantum chromodynamics, 158, 220, 222
atoms, properties of, 100–101, 111, 115, 127–128, 141, 265

Bartel, Norbert, 40
baryon number, 143
baryons, 8, 117, 131, 136, 143, 145, 147–148, 155–156, 157 (Fig. 1)
charmed, 161, 162 (Fig. 3)
beauty quark. *See* quark, bottom
beta decay, 95, 101, 118, 136, 153, 224 (Fig. 3), 228, 242, 249
Bethe, Hans, 193
Big Bang, 41, 45–47, 59, 67, 71–73, 93, 125, 197–198, 237–238, 255, 258–259
Bjorken, J.D., 137, 154, 165
black holes, 65
Bohr, Niels, 179
Bose-Einstein statistics, 143, 148
bosons, 8, 143, 253
Brahe, Tycho, 31–32
Bruno, Giordano, 74
bubble chambers, 11
Burnell, Jocelyn Bell, 89

3C 273, 66–67
3C 144 (Crab Nebula), 66
3C 405 (Cygnus A), 66
3C 274 (M87), 66 (Fig. 3)
Calisto, 33
Carton, Eli, 258
celestial sphere, 28
CERN (European Center for Nuclear Research), 8, 97, 120, 266
CHAMPS (charmed massive particles), 196–197
charm, 7, 12, 127, 137–139, 141, 154, 160, 267
charmonium, 7, 138, 156, 158, 159 (Fig. 2), 160, 176. *See also* J/psi particle
Cicero, 30
Cline, David B., 161
Cocconi, Giuseppe, 74, 89
color, 8, 125, 135–136, 141–142, 149–154, 218–220, 230, 241, 244
conservation laws, 143
constellations (stars), 27
Copernican system, 30–31
Copernicus, Nicolaus, 30–31
cosmic background radiation (3 degree), 71, 93–94
cosmic rays, 123–124, 168
cosmological principle, 41, 68
cosmology, 93, 109, 171, 192–193, 196, 239, 252–253
Costa, Peter, 16
coupling constants, 204, 216–217, 224, 229, 233 (Fig. 5), 234
Crab Nebula (M1 or NGC 1952), 66
Cronin, James W., 259

Crystal Ball detector, 15
Curie, Madame, 169
Cygnus A, 63, 64 (Fig. 2)

Dalton, John, 101, 127, 173
dark matter, 180, 196–197
Davis, Raymond, 194–195
days of the week, 28, 29 (Table 1)
dimensions, unit of, 50
dinosaurs, extinction of, 36, 291
Dirac, Paul Adrien Maurice, 13, 129, 261
distance, logarithmic scale for, 52, 53 (Fig. 6)
Drake, Frank, 91
dysmetria (fear of mathematics), 286, 288–290

early Universe, 47
Earth
  age of, 80–81
  antineutrinos from radioactive decay, 169–170
  atmosphere, 80–81
  heating by radioactive decay, 79–80, 169
  volcanos, 80
eccentric motion, 29–30
education
  mathematics, 288–290
  physics, teaching of, 19–20, 111, 286
  science, 285–287
Egyptian gods, 60
eightfold way, 102–103, 112, 115, 116 (Fig. 4), 127, 133, 144–146, 148, 155–156, 157 (Fig. 1), 164, 175, 186
Einstein, Albert, 19, 21, 65, 240, 245, 248
electromagnetic radiation, 48 (Fig. 4), 49, 129
  penetration of the Earth's atmosphere, 61 (Fig. 1)
  radio waves, 61
electromagnetism, 60, 97 (Table 2), 99, 119, 127, 135–136, 142, 151, 153, 239–240, 248
electron neutrino. *See* neutrino, electron
electron-positron annihilation, 7
electrons, 13, 94, 99–101, 115, 118, 122, 127, 129–131, 136
electroweak theory, 8, 15, 98, 112, 120, 122, 125, 136–137, 153, 180, 189, 203–212, 242, 266
  history of, 205–210
  renormalizability, 207–208
  selection rules, 211
elementary-particle physics, 93, 104, 109, 126–127, 171–173, 192–193, 239, 252–253, 286–287
elementary particles, 181, 189
  classification of, 131 (Fig. 1)
  definitions of, 132, 133 (Table 1), 267 (Table 1)
  number of, 121, 186, 187 (Fig. 4)
  periodic Table of, 188 (Fig. 5)
elements, number of known, 183 (Fig. 1), 184
energy resources, 86–87
English words of astrological origin, 27–28
epicycles, 29–30
equants, 29–30
erosion, 85
Ewen, Harold, 91
extrasolar planetary systems, 78
extraterrestrial life
  communication with, 90–92
  likely stars for, 77–78, 82
  search for, 77, 83, 89, 91–92
  survival time, 88

Fermi-Dirac statistics, 143, 148
Fermilab, 265
fermions, 103, 143, 253
Finnegans Wake (Joyce), 134
Fitch, Val L., 259
flavor, 141, 151, 153, 241
force particles (carriers of), 97 (Table 2), 257 (Table 1)
fractions, 56

Galactic Center, radio radiation, 63
galaxies
  distance, 40
  distribution of, 68, 69 (Fig. 4), 70 (Fig. 5)
  starburst, 78
Galileo, 33–34, 36
gallium, 113
gamma decay, 102
Gamow, George, 71, 99–100, 115, 174–175, 267–268
Ganymede, 33
gauge interactions, 151, 204
gauge invariance, 217–218, 227, 232–233
gauge theory, 8
Gell-Mann, Murray, 8, 98, 102, 115–117, 134, 144, 146, 150, 164–165, 175, 186
George III, King of England, 35
germanium, 113
Gilbert, Walter, 19

Glashow, Sheldon L., 3–23 (*biography*), 16–23, 97, 154
gluons, 45–46, 97 (Table 2), 120, 127, 136, 151–153, 218–219, 230, 242, 249–250
Goldhaber, George, 139, 167
grand unified theory (GUT), 111, 204, 242–247, 251, 257–261
gravitons, 97 (Table 2)
gravity, 97 (Table 2), 109, 136, 151, 193, 248, 253–254, 261
Greek cosmology, 27–28
Greenberg, Oscar W., 149
greenhouse effect, 59–60, 79
Gross, David, 222

hadrons, 117, 131, 133–135, 137, 139, 142–143, 146, 150–151, 156, 159–161, 175–176, 209–210, 219
  charmed, 138–139, 250
  number of, 185 (Fig. 3)
  supermultiplets, 144, 148, 155–157 (Fig. 1)
Hazard, Cyril, 66
Heisenberg, Werner, 19
heliocentric theory, 30–31, 34
helium, 45, 87
  abundance in the universe, 96–97
Herschel, William, 35
Hertz, Heinrich, 60
Higgs boson, 121, 186
high-energy physics. *See* elementary-particle physics
Horowitz, Paul, 76, 91
Hubble, Edwin, 37–40
Hubble constant, 39–40
Hubble law, 38, 39 (Fig. 4)
Huyghens, Christian, 74
hydrogen, 128
  21-cm radiation, 90–91
  abundance in the universe, 96
  hyperfine frequency of, 90

Iliopoulos, John, 137, 154, 166
integers, 55–57
intermediate vector bosons, 8, 136, 153. *See also* W and Z particles
International Committee on Future Accelerators, 275
International Conference on High Energy Physics (Leipzig, 1984), 15
Io, 33
IRAS (Infrared Astronomy Satellite), 78
irrational numbers, 56–57
island universes, 37
isotopes, 111–112
  number of, 184 (Fig. 2), 185
isotopic spin, 145, 147, 175
  multiplets, 144

J/psi particle, 7, 98, 138, 142, 155–156, 158–160, 166–167, 176, 212, 266–267. *See also* mesons
Jansky, Karl, 63
Jupiter, 33–35

Kant, Immanuel, 37, 79
kaon. *See* meson, K
Kelvin, (Lord) William Thompson, 182
Kepler, Johannes, 31–32, 36
Kepler's Laws of Planetary Motion, 31–32, 34
Kepler's supernova, 33
Kogut, John, 151

lambda particle, 145, 147, 162 (Fig. 3), 165
Lavoisier, Antoine Laurent, 100–101, 121
lead isotopes, 80
Lederman, Leon, 19, 118
lepton-quark symmetry, 177
leptons, 94 (Table 1), 100, 103, 117–118, 131, 137, 142–143, 153, 165–166, 176, 217, 222–223, 229, 250, 257, 267
Leverrier, Urbain, 35
Lie groups, 144, 258
light, 48–49
lithium, 45
Little Green Men, 90
Local Group (of galaxies), 69
logarithmic scale, 44–45
Lowell, Percival, 36, 76

M5, 66
M101, 40
M82 (3C 231), 66 (Fig. 3)
M87 (3C 274), 66 (Fig. 3)
M1 (Crab Nebula), 66
magnetic moments, 207
magnetic monopoles, 13, 232, 246, 261
Maiani, Luciano, 137, 154, 166
Malthus, Thomas Robert, 85
Manhattan Project, 22
Mann, Alfred, 161
Marconi, Guglielmo, 61–62
Mars, 35, 76, 79
matter-antimatter asymmetry, 95–96

Maxwell, James Clerk, 13, 60, 99, 239, 241, 248
mechanics, Newtonian, 34–36, 99
Mendeleev, Dmitri Ivanovich, 111–113, 173–174
Mercury, 35
mesons, 112, 117, 123–124, 130–131, 135, 143, 145, 149, 155, 160, 165
  K, 109, 148, 154, 210
  phi, 158
  pi, 130–131, 133, 149–150, 162, 168, 175, 268
  upsilon, 250, 265–267
metric system, 51, 52 (prefixes of, Table 3)
Metrodorus, 74
Milky Way, 27, 33, 36, 59, 68–69, 79
Miller, Stanley L., 81
Miller–Urey experiment, 81
mixing angles, 215, 229, 257, 260
Moon, 29, 33
Morrison, Philip, 74, 89
Moseley, Henry, 104, 265
muon neutrino. *See* neutrinos, muon
muons, 94, 109, 118, 124, 142, 150, 162–163, 168, 251
mutual assured destruction (MAD), 292–293

natural resources, exhaustion of, 86–87
Ne'eman, Yuval, 102, 115, 117, 144, 150, 164, 186
Nemesis, 12, 36
Neptune, 35–36
neptunium, 36
neutral currents, 120, 154, 206, 208–210, 226, 228–229, 242, 249, 259
neutrino astronomy, 168
neutrinos, 97, 99, 109, 115, 118, 122, 124, 131, 138, 153–154, 161, 163, 168, 177–178, 195–196, 223, 228, 230, 242, 251, 260
  electron, 94, 118, 142, 163, 166
  mass, 211, 246
  muon, 94, 118, 142, 154, 163, 166
  oscillation, 211, 246
  spin, 222
  from supernova 1987a, 196
  tau, 94
neutrons, 102, 112, 115, 117, 128–129, 133–134, 142
New Galactic Catalog (NGC), 66
Newton, Isaac, 32, 34, 127
NGC 1952 (Crab Nebula), 66
Nicetas, 30
Nobel Prize in Physics, 21–22, 72
Nobel Prizes (American), 285
nonluminous matter. *See* dark matter
nuclear disarmament, 291–293
nuclear holocaust, 83–84, 291–293
nuclear physics, 14, 102, 114
nucleons, 99, 103, 131
nucleus, atomic, 111
numbers, 49, 50 (Fig. 5)

Oblers' paradox, 41
omega minus particle, 116, 124, 148, 165, 175, 186
One, Two, Three, Infinity (Gamow, 1947), 99, 115
overpopulation, 85

parity violation, 124, 189, 228
particle physics. *See* elementary-particle physics
Pauli, Wolfgang, 131, 143–144, 168
Pauli exclusion principle, 131, 143–144, 148
Peebles, P.J.E., 72
Penzias, Arno, 59, 72
periodic Table, 101–102, 111–112, 114
Perl, Martin, 119
photons, 97 (Table 2), 104, 120, 129–130, 135, 151, 230, 249–250
physics, Federal budget for, 282 (Fig. 5)
pi, 49, 57
pions. *See* mesons, pi
Planck length, 179, 253–255
planetary motions, 28–36, 32 (speed *vs.* distance from Sun, Fig. 2)
planets, 29–30
Plutarch, 30
Pluto, 36, 76
plutonium, 36
Politzer, David H., 156, 222
pollution, 84
population growth, 85, 86 (Fig.1)
positrons, 123, 129, 230
powers of ten notation, 49
Project Ozma (SETI), 91
protons, 102, 112, 115, 117–118, 123, 128–130, 134, 136, 142–143, 251
  decay, 168, 204, 211, 215, 234–235, 236 (Fig. 6), 244–245, 259–260
protoplanetary disks, 78
psi particle (mesons). *See* J/psi particle. *See also* mesons
Ptolemaic system, 29
pulsars, 89

Purcell, Edward, 13, 90–91

quantum chromodynamics, 8, 15, 98, 103, 112, 120, 122, 135, 176, 180–182, 204, 214–216, 218–220, 219 (Fig. 1), 241, 243, 266
asymptotic freedom, 220–222
vacuum polarization, 221–222, 223 (Fig. 2)
quantum electrodynamics, 120, 129, 203, 215–217, 219 (Fig. 1), 240–241
quantum field theory, 240
quantum gravity, 180–182, 253, 255
quantum mechanics, 99, 101, 109, 114, 139, 240
quantum numbers, 143
quark model, 8, 138
quarks, 8, 11, 13, 18, 21, 45–46, 94, 94 (Table 1), 98, 100, 103, 112, 117–119, 122, 124, 127, 134–135, 137, 139, 141–142, 144, 146–149, 151–156, 161, 164–167, 175–178, 181, 209, 212, 214, 217–220, 222–224, 228–232, 235, 241, 252–253, 257, 261
beauty, 119, 124
bottom, 119, 250, 256, 260
charmed, 12, 94–95, 119, 124, 137, 148, 154, 156, 160–161, 163–167, 177, 209, 250
color, 161
flavor, 161, 163
fractional charge, 219, 230, 232
origin of name, 134–135, 165
strange, 118, 134, 177, 250
top, 103, 119, 250, 256, 260
quasars, 59, 66–68

Rabi, Isadore, 118, 181
Radar (Radio Detection and Ranging), 63
radio astronomy, 63, 91
radio galaxies, 63–67
Cygnus A, 63, 64 (Fig. 2)
M82, 66 (Fig. 3)
M87, 66 (Fig. 3)
radio waves, 62 (Table 2)
radioactivity, 95, 128, 169
dating by, 79–81, 169
decay, 79–80, 169
rational numbers, 56
Reagan, Ronald, 22
red shift (galaxies), 39–40
reductio ad absurdum, 56–57
relativity, 99, 109, 140, 240
renormalizability, 207–208
Richter, Burton Jr., 155, 166
Roman Catholic Church, 30–31
Rubbia, Carlo, 97, 120, 122, 124, 161
Rutherford, Ernest, 101, 105, 141, 169, 174

Sakharov, Andrei, 96, 259
Salam, Abdus, 22, 97, 153, 249
salt, symmetries of, 226–228
Samios, Nicholas P., 116, 161, 167
Saturn, 35
scandium, 113
Schmidt, Maarten, 67
Schwartz, Mel, 118
Schwinger, Julian, 153
scientists, responsibility for weapons development, 291
SETI (Search for Extraterrestrial Intelligence), 91
Shapiro, Irwin, 40
Sizzi, Francesco, 34
SPEAR (Stanford electron-positron collider), 8, 119, 138–139, 166
special unitary groups. *See* SU groups
speeds, logarithmic scale for, 52, 53 (Fig. 7), 54
spin angular momentum, 143, 147–148
spontaneous symmetry breakdown. *See also* SU groups and grand unified theory, 226–229, 232, 234
stages of life, 42 (Table 1), 43 (Fig. 1), 44 (Fig. 2), 45
standard model. *See also* grand unified theory, 26, 98, 125, 180, 203–204, 256, 273 (Table 1)
Stanford Linear Accelerator Center (SLAC), 8, 12, 266
star motions, 28 (Fig. 1), 29
Steinberger, Jack, 118
Stoney, George Johnstown, 13
strange particles, 115, 118, 134, 175, 268
strangeness, 127, 145, 147–148
strong force, 97 (Table 2), 120, 130, 136, 142, 248, 252–254, 266
strong interactions, 8, 151, 176
SU groups, 144, 146, 148, 150, 206–210, 212, 220, 222, 224–235, 237–238, 244–245, 258
fundamental particles, 231 (Table 1)
subatomic physics, 17, 21
Sun, 193–194
energy generation, 193
evolution of, 193
neutrinos from, 194–195
Superconducting Super Collider, 11, 22, 105,

125, 198, 265–284
challenges of, 271–272
cost, 281–284, 282 (Fig. 5)
design, 275–279, 276 (Fig. 1), 277 (Fig. 2), 278 (*Reference Design Study*)
international cooperation on, 280
magnets, 278 (Fig. 3), 278–280, 279 (Fig. 4)
reasons for building, 268–271, 273–275
spinoff from, 272–273
superheavy atoms, 261
supermultiplets. *See also* hadrons, 144, 148, 155–156, 157 (Fig. 1)
supernova, 33, 82
1987a, 195–196
superstrings, 182–183, 255–256
survival time for intelligent species, 88
Susskind, Leonard, 151

t' Hooft, Gerhard, 153
Tang, 11
taons. *See* tau particles
tau leptons, 94, 119, 124, 163, 212, 251, 256, 266–267
tau neutrino. *See* neutrino, tau
Teflon, 11, 22
Tevatron II, 265
Theory of Everything, 190, 255–256
thermodynamics, second law of, 95
Third Cambridge Catalog (3C Catalog), 66
thorium, 79, 82
time, arrow of, 259
time reversal, 95–96
Ting, Samuel C.C., 155, 166
transcendental numbers, 57–58
truth quark. *See* quark, top
Tycho's supernova, 33

unified field theory, 248–251
unitary symmetry. *See also* SU groups, 209
units of length, 50, 51 (Table 2)
Universe
Age of Atoms, 46
Age of Ions, 46
Age of Leptons, 45
Age of Nucleons and Antinucleons, 45
Age of Nucleosynthesis, 45–46
Age of Quarks and Gluons, 45
Age of Stars and Galaxies, 46
evolution, (Fig. 3), 45–47
expansion, 39–40, 93
number of protons, 49–50
symmetry breaking, 198, 258
upsilon particles. *See* mesons, upsilon
uranium, 22, 35, 79, 82
Uranus, 35–36
Urey, Harold, 81

van de Hulst, H.C., 91
Van der Meer, Simon, 97
vector bosons, 139, 151, 153
Venice, 33
Venus, 35, 76, 79
Virgo Cluster, 69, 73
Virgo Supercluster, 69
volcanos
origin of Earth atmosphere, 80
outgassing from, 80–81

W particles, 8, 97 (Table 2), 98, 105, 111, 120, 122, 124–125, 153–154, 224, 226–228, 230, 242, 249–250, 253, 260
weak force, 97 (Table 2), 136, 142, 166, 252–254
weak interactions, 153, 222–224, 226 (Fig. 4), 242, 250
Weinberg, Steven, 22, 97, 153, 249
Wilczek, Frank, 222
Wilkinson, Denys, 176
Wilson, Kenneth, 151
Wilson, Robert W., 59, 72
WIMPS (weakly interacting massive particles), 196
women, in physics, 23
Wright, Thomas, 79
Wu, C.S., 124

X rays, 104
xi particles, 116 (Fig. 4), 165

yttrium, 113
Yukawa, Hideki, 123, 130, 153, 175

Z particles, 8, 97 (Table 2), 98, 105, 111, 120, 122, 124–125, 153, 163, 209, 211, 226–228, 230, 249, 253, 260, 267
zeta particle, 13
zirconium, 113
Zweig, George, 134, 146, 165, 175
Zweig's rule, 156, 158

About the Author

One of the most creative theorists of our age, Sheldon Lee Glashow shared the Nobel Prize in 1979 for his pioneering achievements in establishing a key concept in modern physics—the electroweak theory, a concept that successfully integrated electromagnetic and weak forces.

Later, he introduced the idea of a fourth quark, christened "charm," which was subsequently confirmed experimentally. Professor Glashow is also celebrated for his contributions to the search for a Grand Unified Theory, linking all of the forces of nature with the strong force.

The author of more than 200 publications, his autobiographical *Interactions: A Journey Through the Mind of a Particle Physicist*, written together with Ben Bova, appeared in 1988. A frequent contributor to the popular scientific literature, Professor Glashow is the physics editor of the new student mathematics and science magazine, *Quantum*.

Born in New York City, Dr. Glashow is the son of immigrant parents who fled czarist Russia at the turn of the century. He graduated from the illustrious Bronx High School of Science and went on to earn his undergraduate degree at Cornell and his doctorate at Harvard.

Dr. Glashow returned to Harvard in the mid-sixties as a professor of physics, after teaching and performing research at CERN's Niels Bohr Institute, Caltech, Stanford, and Berkeley. An innovative teacher of science, he now holds the Higgins and Mellon chairs at Harvard.

Elected to the National Academy of Sciences and the American Academy of Arts and Sciences, Professor Glashow has received numerous other honors, including the Oppenheimer Prize and five honorary doctoral degrees from universities in the U.S., France, and Israel.